Male hypogonadism

UNI-MED

Prof. Dr. Friedrich Jockenhövel
Chefarzt der Medizinischen Klinik
Evangelisches Krankenhaus Herne
Wiescherstr. 24
D-44623 Herne
Germany

Dr. Markus Schubert
Oberarzt der Klinik II und Poliklinik für Innere Medizin
Klinikum der Universität zu Köln
Kerpener Str. 62
D-50937 Köln
Germany

Jockenhövel, Friedrich/Schubert, Markus:
Male hypogonadism/Friedrich Jockenhövel and Markus Schubert.-
2nd edition - Bremen: UNI-MED, 2007
(UNI-MED SCIENCE)
ISBN 978-3-89599-859-1

Printed in Europe

MEDICINE - STATE OF THE ART

UNI-MED Verlag AG, one of the leading medical publishing companies in Germany, presents its highly successful series of scientific textbooks, covering all medical subjects. The authors are specialists in their fields and present the topics precisely, comprehensively, and with the facility of quick reference in mind. The books will be most useful for all doctors who wish to keep up to date with the latest developments in medicine.

Foreword

The aim of this book is to provide cutting edge knowledge on the diagnostics and therapeutic procedure for men with impaired testicular endocrine function.

The book provides in depth background information on anatomy and physiology of endocrine testicular function. Where necessary, the most recent developments in molecular biology are covered. The clinical presentation of male hypogonadism is described in detail, as well as all procedures necessary for a precise diagnosis.

An extensive chapter deals with all aspects of androgen substitution therapy and the misuse of androgens. Most recent and future developments, such as transdermal applications including testosterone gel and the long-acting intramuscular testosterone formulations are described in depth. For each formulation and application system the individual advantages and disadvantages are discussed.

Androgen deficiency in the aging male is covered by a separate chapter. This recognizes the importance of this growing field of interest and the specific diagnostic and therapeutic aspects of this subject. Also, a specialised chapter covers the subject of estrogen deficiency and therapy with estrogens in men.

We are most grateful to the staff from UNI-MED Verlag for their excellent support.

Friedrich Jockenhövel
Markus Schubert

Preface

This excellent text on male hypogonadism provides a comprehensive yet focused view of testicular function and hormone related abnormalities involving the male reproductive system. Friedrich Jockenhövel has managed to provide the background that is so often demanded by the specialist and presents the information in a fashion that is easily understandable by the generalist.

There has been remarkable progress in male reproductive medicine in the past several decades. The book demonstrates how the practitioner distinguishes central (hypothalamic or pituitary) from primary testicular causes of testosterone deficiency and provides a differential diagnostic approach to congenital and acquired forms of hypogonadism. The complex mechanisms of action of androgens at multiple target organs have provided the groundwork for understanding the benefits and risk of testosterone therapy. A perception of male hormones as essential for male sexuality has now been greatly expanded to a broader understanding of the metabolic aspects of circulating androgens. This text describes the role of androgens on such diverse organs as liver, bone, muscle, fat and the central nervous system. We have come to know testosterone as not only a hormone, but a pre-hormone serving as the precursor to metabolically active hormones (dihydrotestosterone and estradiol). Thus, testosterone affects may be mediated through both androgen and estrogen receptors. Androgen deficiency presents differently at different ages. This book characterizes testosterone deficiency occurring during fetal development at puberty and adulthood. We have come to recognize that testosterone levels in the blood fall progressively with age and that there is a significant percentage of the older male population whose circulating testosterone levels are considerably lower than that of their young healthy male counterparts. We have also come to understand the effects of the increasing sex hormone binding globulin concentration and its impact on decreasing free and bioavailable testosterone as men get older. The next decade will undoubtedly reveal the relative benefits and risks of treating older males with androgen deficiency with therapeutic doses of testosterone. The recognition of significant bone disease in men with aromatase and estrogen receptor deficiency has lead us to appreciate the important role of estrogens in normal male physiology. This recognition will influence the way we look at new synthetic androgens with selective metabolic actions. Multiple new androgenic preparations have become available giving greater flexibility to the physician in treating hypogonadal disorders. At present, transdermal preparations of testosterone have captured much of the market in the United States and Europe, and the future will reveal new safe oral testosterone delivery systems and long acting depo preparations requiring only three to four injections per year. In the next several years, large numbers of synthetic androgen receptor modulators will also become available. These drugs will have selective effects on various target organs and allow focused therapy with androgens for specific indications. This text also describes condition of estrogen deficiency in the male, originally appreciated by the careful scrutiny of a few individuals who had estrogen receptor defects and/or the inability to convert androgens to estrogens due to aromatase deficiencies. In the latter sections of the book, Dr. Jockenhövel's case management format is ideally suited for the clinician in his/her strategic approach to the diagnosis and management of the androgen deficient patient.

Ronald S. Swerdloff, M.D.
Professor of Medicine
David Geffen UCLA School of Medicine
Chief, Division of Endocrinology
Harbor-UCLA Medical Center
Director of Mellon, WHO and NIH Centers of Excellence in Male Reproduction

Contents

Anatomy and physiology of the testis

1. Anatomy and physiology of the testis

The dual function of the testis as an endocrine (hormone-producing) and exocrine (sperm-producing) gland affords it a special position among the endocrine organs. This is reflected in the anatomy and physiology of the testis. The testicular parenchyma, which is surrounded by a solid capsule (tunica albuginea), consists of seminiferous tubules in which gametes are produced (spermatogenesis). Septa of connective tissue divide the testis into 200-300 lobules; these combine in the rete testis. Each lobule contains two to three seminiferous tubules which average 50 cm in length and are very convoluted. Each testis contains altogether 600-900 seminiferous tubules, approximately 350 m in total length. This explains the enormous reproductive capacity of the human male, who can produce approximately 10-20 million gametes per day. In the interstices between the seminiferous tubules are the steroid-producing Leydig cells, as well as the necessary blood and lymph vessels and nerves. Approximately 85-90 % of the volume of the testis is taken up by the seminiferous tubules, and only 10-15 % by the interstices (☞ Figure 1.1)[209].

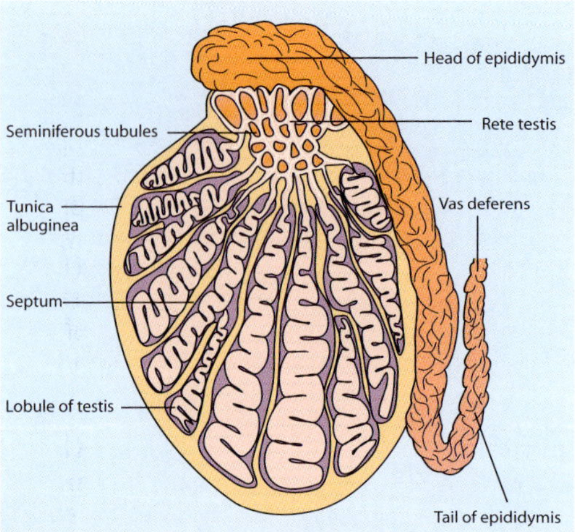

a

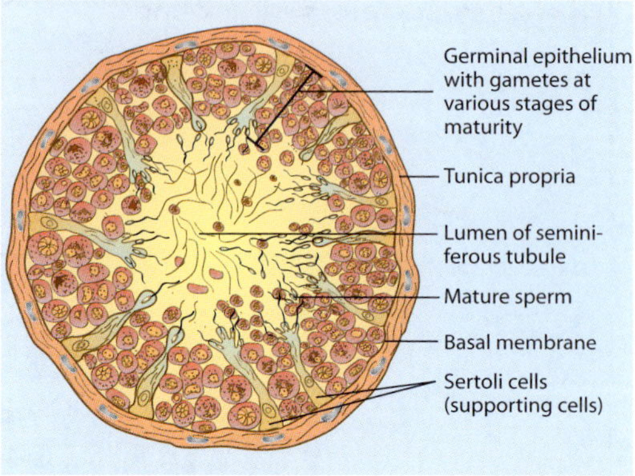

b

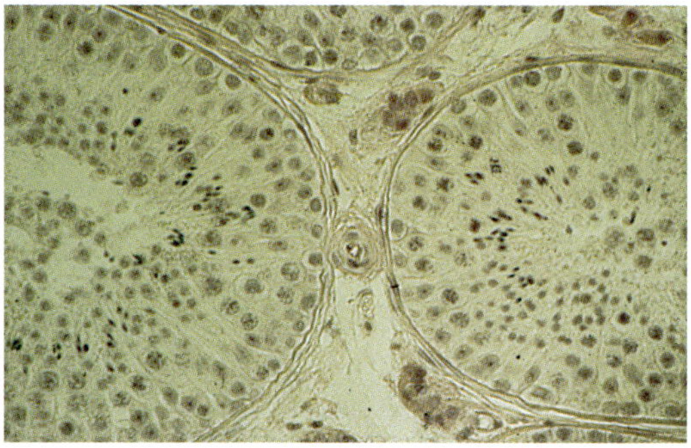

c

Figure 1.1: **(a)** Macroscopic cross-section through the testis and epididymis; **(b)** microscopic section from (a) showing seminiferous tubules; **(c)** histological cross section showing cut surfaces of three seminiferous tubules with intact spermatogenesis. Between the tubules is the interstice containing Leydig cells and a blood vessel (HE stain, 40x magnification).

1.1. Hypothalamic-hypophyseal regulation

The hypothalamus controls the function of the testis by means of the pituitary hormones, luteinizing hormone (LH) and follicle stimulating hormone (FSH) (☞ Figure 1.2). In the hypothalamus, gonadotropin releasing hormone (GnRH) is produced under the stimulating and inhibiting influence of neurotransmitters. GnRH promotes the production and release of the gonadotropins LH and FSH in the hypophysis.

In humans, GnRH neurons are initially laid down in the region of the cribriform plate, and adhesion

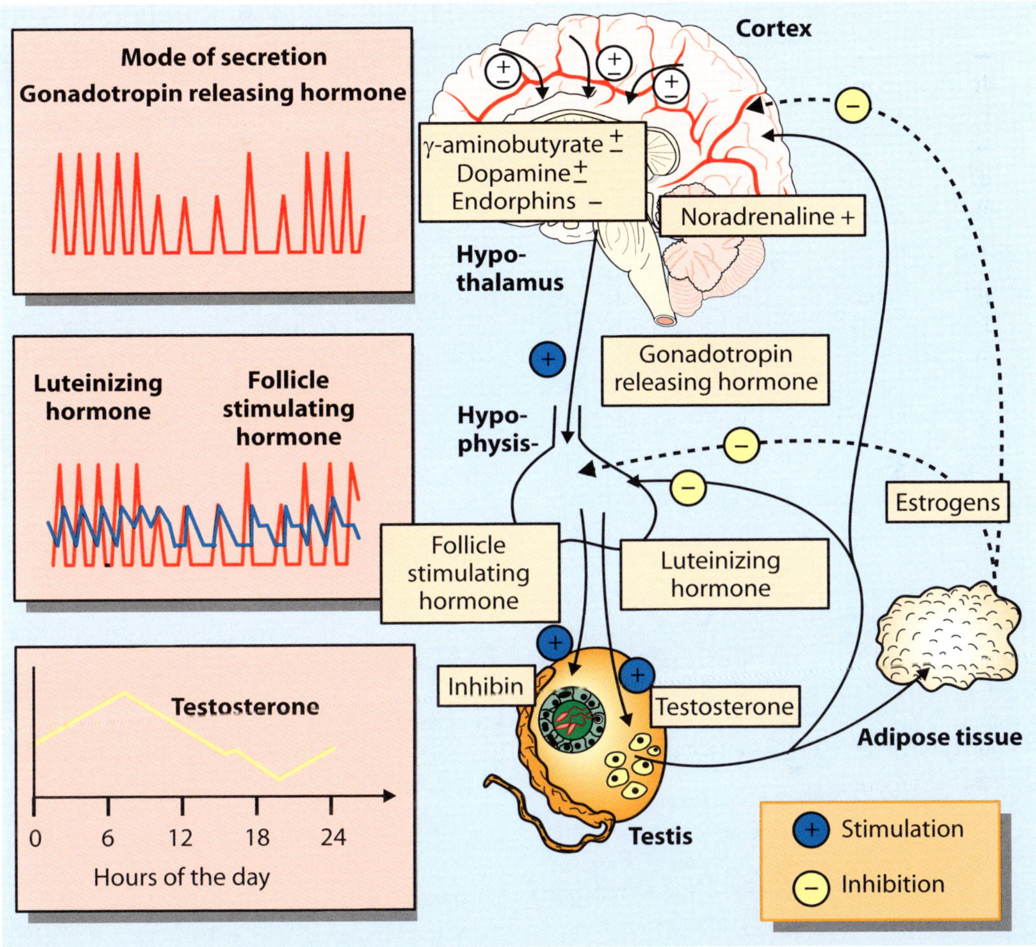

Figure 1.2: Regulation of testicular function and secretion pattern of the hormones involved. Gonadotropin releasing hormone (GnRH) is secreted from the hypothalamus under the influence of neurotransmitters (γ-aminobutyrate, catecholamines, endorphins) and from the cortex in a pulsatile fashion, and stimulates production of the gonadotropins luteinizing hormone (LH) and follicle stimulating hormone (FSH) in the hypophysis. LH (shown in red) follows the secretion profile of GnRH very closely, whereas FSH (blue) does not display such clear pulsatility owing to its longer half-life. LH stimulates testosterone biosynthesis in the Leydig cells. Testosterone exerts a direct, and after metabolism to estrogens an indirect, negative feedback effect on the secretion of GnRH and gonadotropins. Together with testosterone, FSH stimulates the production of peptides such as inhibin, which in turn exerts a negative feedback effect on FSH.

molecules cause them to migrate to the arcuate nucleus and mediobasal brain centers of the hypothalamus. Already in the fetus GnRH is produced under the control of neurotransmitters. Dopamine, serotonin and GABA exert inhibitory effects and NPY stimulating effects.

GnRH is formed in the arcuate nucleus and the preoptic region of the hypothalamus. Its axons innervate the median eminence and secrete GnRH into the portal system of the hypophysis. GnRH is initially produced in the form of prepro GnRH. Prepro GnRH is encoded by a gene on chromosome 8 (8p21-p11.2). The decapeptide GnRH arises by post-translational enzymatic cleavage of the leading 23 amino acids ('pre' portion) and of the supporting 56 amino acids ('pro' portion = GnRH-associated peptide, GAP). The enzyme prohormone convertase-1 is responsible for this.

GnRH is secreted in regular pulses, with peaks every 90-120 min. This is due to the intrinsic ability of GnRH neurons to secrete episodically. This basic rhythm is modulated by numerous neurotransmitters. Pulse amplitude and pulse frequency affect the amount of LH and FSH secreted by the hypophysis. α-Adrenergic impulses have a stimulatory action, and β-adrenergic and dopamine impulses an inhibitory action on GnRH secretion. Endorphins display cycle-dependent effects in women but have a predominantly inhibitory action in men. Testosterone and progesterone slow down this pulse rate, presumably mediated by β-endorphins. The adverse effect of stress on reproductive function is well known. Several factors are involved: corticotropin releasing hormone (CRH) inhibits GnRH secretion through direct neuronal contact between the paraventricular nucleus and preoptic region. The level of prolactin, which is often raised in stress, further reduces the GnRH pulse rate. In addition, central cytokines, such as the inhibitory interleukin 1, also appear to play a role. Recent findings indicate that leptin, presumably indirectly via neuropeptide Y (NPY), intervenes in the feedback mechanism. Therefore, the administration of leptin to leptin-deficient obese infertile rats not only reduces weight but also restores fertility. Leptin is also likely to be involved in the initiation of puberty. GnRH has a half-life of 5-10 min but cannot be measured owing to its low systemic concentrations.

In the anterior lobe of the pituitary, GnRH binds to specific receptor (typical G-protein-bound receptors with seven transmembranous loops) on the gonadotropic cells and initiates the gene expression of α- and β-chains of FSH and LH and their secretion by induction of inositol-1,4,5,-triphosphate, mobilization of intracellular calcium and increased calcium influx. Whereas androgens reduce the number of GnRH receptors, the influence of estrogens is dependent on the concentration of estrogen, the dynamics of the concentration change and other factors.

The pulsatility of GnRH is essential to its gonadotropin-stimulating effect. Continuous administration of GnRH causes gonadotropin production to cease completely owing to down regulation of the GnRH receptor. This effect is utilized therapeutically in the administration of slow-acting GnRH analogs in conditions which are dependent on sex steroids (e.g. endometriosis, carcinoma of the prostate, precocious puberty). In patients with hypothalamic-induced hypogonadism, e.g. Kallmann's syndrome, correct pulsatile administration of GnRH can establish normal LH and FSH concentrations and hence normal gonadal function.

The polypeptides LH and FSH are large glycoproteins like thyroid stimulating hormone (or thyrotropin, TSH) and human chorionic gonadotropin (hCG). A common characteristic is the non-covalent binding of two peptide chains (α and β) to a heterodimer; the α-chain is identical in all four hormones, and therefore the biological effect is mediated by the β-chain. The α- and β-chains are encoded by genes on different chromosomes. The α-chain is encoded by a gene on chromosome 6 (6q12.21), FSH-β on chromosome 11 (11p13) and LH-β and hCG-β on chromosome 19 (19q13.32). Isolated subunits and homodimers show no biological activity. The strong similarity of the basic structure, the high homology of the amino acid sequence and the organizational structure of the genes make the phylogenetic origin of all glycoproteins by gene duplication from a common precursor gene very likely. In phylogenesis, the gonadotropins developed later than TSH; they show more homology among themselves than with TSH. Gene duplication of LH and hCG occurs even later, so these hormones are almost identical and have the same biological action. However, hCG has a significantly longer half-life. Owing to gene duplication, which is still young in terms of evolutionary development, the genes for the β-chains of LH and hCG lie in a gene complex with six gene duplications of hCG-β and one gene for LH-β on chromosome 19 (19q13.32).

Moreover, a common feature of all chains of the glycoprotein hormone family is co-translational glycosylation. This means that, at certain points on the peptide chain, carbohydrates are bound to the amino acid asparagine. The structure is usually Glc_{1-3}(α-mannose 4-6-mannose β_{1-4}Glc)nAC β_{1-4}Glc-NAc-Asn. These carbohydrates are then further modified by the removal of glucose and mannose portions and the incorporation of other oligosaccharides (e.g. fucose, galactose and sialic acid), so that even more complex carbohydrate side-chains arise on the amino acid chain. This process is not always identical. Therefore, minimal differences in carbohydrate side-chains cause different forms of glycoproteins to be secreted, which may be separated from each other by special procedures (e.g. isoelectric focusing). Hence, seven different species of LH are found in man which differ only in their carbohydrate composition. The carbohydrate side-chains influence the tertiary structure, binding of α- and β-chains, half-life in the systemic circulation, binding to specific receptors and also intracellular signal transduction after receptor binding in the target cell. In particular, a high concentration of sialic acid prevents metabolism of glycoproteins in the liver and kidneys, and thus prolongs the half-life and biological effect. In TSH and FSH, the carbohydrate side-chains terminate with sialic acid, which explains the longer half-life compared with LH. Thus, LH has a half-life of only 20 min, whereas that of FSH is 3 h and hCG 5 h[181].

LH binds to specific receptors on the surface of the Leydig cells and, mediated by cyclic adenosine monophosphate (cAMP), causes an increase in intracellular cholesterol and increased gene expression of enzymes of steroid production, in particular the key enzyme 20,22-desmolase. This initiates biosynthesis of testosterone with cleavage of the side-chain from cholesterol. The feedback control of LH production in man occurs via testosterone and its metabolite estradiol. Testosterone has an inhibitory effect on the neurons producing GnRH, and exerts only a slight suppressant effect on hypophyseal LH production. In contrast, estradiol has an inhibitory action on the hypophysis and hypothalamus.

FSH binds to receptors on the Sertoli cells and promotes spermatogenesis in a manner that is as yet unexplained. Among other things, the activity of the enzyme aromatase, which converts androgens to estrogens, is stimulated in the Sertoli cells. In addition to a number of other proteins, the hormones inhibin and activin are formed in the Sertoli cells under the influence of FSH. Inhibin is a heterodimer of the α and β subunits; there are two β-variants (β_a and β_b). Hetero- and homodimers of the β-chains are called activin A (β_a-β_b) and activin B (β_b-β_b). Inhibin is an important component in the feedback system controlling FSH secretion. In isolated functional disturbances of the Sertoli cells (e.g. Sertoli-cell-only syndrome or following radiotherapy or chemotherapy), inhibin deficiency is indicated by a sharp rise in FSH while LH remains at normal levels. The physiological significance of FSH-stimulating activins has not been conclusively explained; however, they appear to be less important. Testosterone and estrogens also exert a negative feedback effect on FSH via their effect on GnRH.

Gonadotropins, on the other hand, promote testicular function directly. LH binds to specific receptors on the Leydig cells and, in a cAMP-mediated process, stimulates the production of enzymes which initiate biosynthesis of testosterone with cleavage of the side-chain from cholesterol. FSH binds to receptors on the surface of the Sertoli cells and the spermatogonia and, together with testosterone, promotes spermatogenesis. Like the GnRH receptor, gonadotropin receptors are G-protein-bound receptors with seven transmembranous loops. The genes of the LH and FSH receptors both lie on chromosome 2.

1.2. Spermatogenesis

Spermatogenesis is a complicated process of mitotic and meiotic cell divisions in which haploid spermatids are formed from diploid spermatogonia via several intermediate stages (☞ Figure 1.3). Spermatids are transformed into flagellated spermatozoa by a process of metamorphosis, called spermiogenesis. After spermiogenesis is complete, the spermatozoa are released from the germinal epithelium (spermiation) and flow into the epididymis with the fluid from the tubules, where they await ejaculation. The entire process of spermatogenesis and spermiogenesis takes 72 days in humans. Until spermiation, the gametes are dependent on the nursing Sertoli cells, which extend from the basal membrane of the seminiferous tubules deep into the lumen; the Sertoli cells secrete electrolytes and fluid under the influence of FSH and testosterone.

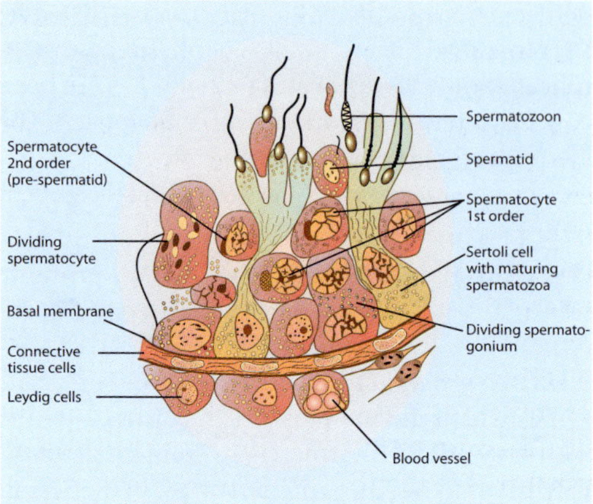

Figure 1.3: Section through a seminiferous tubule. The gametes are embedded between the Sertoli cells and migrate during development from the basal membrane to the lumen, where the spermatozoa are released.

After puberty, spermatogenesis can be maintained without FSH solely by sufficiently high intratesticular concentrations of testosterone. The exact function of the Sertoli cells and the mechanisms behind the paracrine interaction with gametes are unknown; however, the Sertoli cells are imperative to spermatogenesis. At their base, Sertoli cells form 'tight junctions' with one another, which seal the intracellular gap and thus form a blood-testis barrier. Like the blood-brain barrier, the blood-testis barrier is impermeable to macromolecules. The lumen of the seminiferous tubules is divided into two compartments in parallel with the basal membrane: in the basal region in 'front' of the blood-testis barrier are the spermatogonia, and 'behind' the blood-testis barrier all the more advanced stages of spermatogenesis take place. Therefore, the composition of the fluid in the tubules (adluminal compartment) is exclusively determined by secretion from the Sertoli cells, and its composition differs from that of serum or interstitial fluid (e.g. electrolyte, protein, glucose, amino acid content).

In the fetus, the Sertoli cells produce a polypeptide called anti-Müllerian hormone (AMH), which prevents the formation of internal female genitalia (uterus, Fallopian tubes) from the Müllerian duct during sexual differentiation. AMH can be detected in the serum until puberty. The postnatal significance of AMH is unclear. After puberty the production of AMH stops. The Sertoli cells also produce the hormones inhibin and activin. These peptides consist of two polypeptide chains linked together by disulfide bridges; they exert both a positive and a negative feedback effect on FSH[95,176,319,407].

1.3. Testosterone and androgen effect

1.3.1. Testosterone biosynthesis

While the sequence of steroid synthesis including all intermediate stages has been known for decades, new and detailed findings have recently been made with the aid of molecular biology techniques. Hence, the genes of all enzymes required for testosterone biosynthesis have been identified, cloned and expressed in transformed cells. This allows a completely new insight into the regulation of steroidogenesis from a molecular viewpoint, and has shed light on a number of rare disorders caused by enzyme deficiencies in steroid synthesis.

> Testosterone is the most important steroid produced by the testis; 5-7 mg (!) testosterone are produced each day by the Leydig cells of an adult man.

Like all steroid-producing cells, the Leydig cells have a large endoplasmic reticulum and numerous mitochondria. They also typically contain Reinke crystalloids, but the function of these is unknown.

> The parent substance of testosterone biosynthesis is cholesterol, which is mainly synthesized by the Leydig cells; only a small amount of cholesterol is taken from the circulating blood[435]. Cholesterol is stored in form of esters in fat vacuoles in the Leydig cells until further processing. Through a total of five enzymatic stages, cholesterol, which contains 27 carbon atoms, is hydrolyzed to C19-testosterone (☞ Figure 1.4).

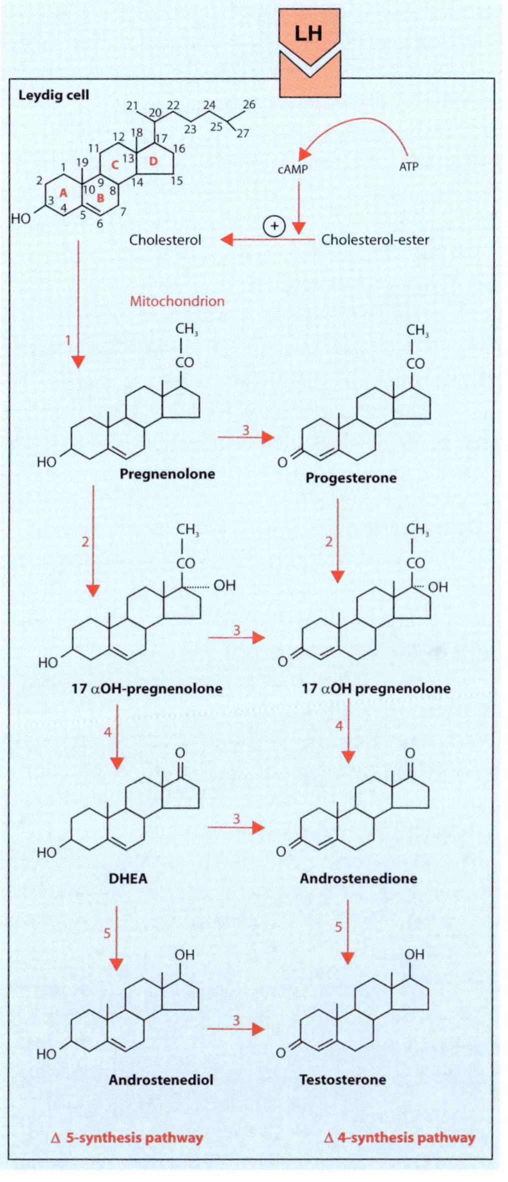

Figure 1.4: Biosynthesis of testosterone in Leydig cells. 1 = 20,22-desmolase; 2 = 17α-hydroxylase; 3 = 3β-hydroxysteroid dehydrogenase; 4 = 17,20-desmolase; 5 = 17β-hydroxysteroid dehydrogenase.

The most significant and rate limiting step in the enzymatic cascade is the conversion of cholesterol to pregnenolone. This takes place on the inner side of the membrane of the mitochondria, where the membrane's own enzyme cytochrome P450scc (side-chain cleavage) catalyzes three consecutive stages: first hydroxylation on atom C20, then on atom C22 and thereafter cleavage between C20 and C22, giving pregnenolone and isocaproic acid. Cytochrome P450ssc, also known as 20,22-desmolase, is the crucial enzyme in all the steroid-producing tissues (adrenal gland, ovary) and is encoded by a gene on chromosome 15. Pregnenolone is the parent substance of all biologically active steroid hormones (corticosteroids, mineralocorticoids, gestagens, estrogens, androgens) and leaves the mitochondria by diffusion. It undergoes further processing in the endoplasmic reticulum. There are two pathways available: these are called the Δ4- or Δ5-synthesis pathways, depending on whether the double bond is located in ring A or ring B. The Δ5-synthesis pathway is preferred in man. Accordingly, hydroxylation frequently occurs first in position C17 by the action of cytochrome P450c17 (17α-hydroxylase), to 17α-hydroxypregnenolone. The weak androgens dehydroepiandrosterone (DHEA) and androstenediol are produced by the enzymes 17,20-desmolase and 17β-hydroxysteroid dehydrogenase. A further important step is the conversion of the less biologically active Δ5-steroids 17α-pregnenolone, DHEA and androstenediol to the corresponding more effective Δ4 steroids 17α-progesterone, androstenedione and testosterone. This step is catalyzed by the enzyme 3β-hydroxysteroid dehydrogenase, and comprises oxidation of the 3β-hydroxyl group to ketone with subsequent transfer of the double bond from C5-C6 on ring B to C4-C5 on ring A (Δ5-Δ4-isomerization). The majority of testosterone produced in this way is immediately released into the blood, and not stored within the testes. Most testosterone is transported by the spermatic vein; a small amount of testosterone is transported in the lymphatic system.

Testosterone biosynthesis is regulated by LH which influences the most critical step of synthesis, the conversion of cholesterol to pregnenolone, via two mechanisms:

- cAMP-mediated stimulation of the synthesis and activity of cytochrome P450ssc
- Protein kinase-C-mediated increase in the production of cholesterol by activation of hydrolase, an ester of cholesterol

Mobilization of cholesterol from the fat vacuoles renders more substrate available to the mitochondria for conversion to pregnenolone. Recent studies indicate that an as yet unidentified factor in the Sertoli cells, which is in turn produced under the influence of FSH, promotes steroid biosynthesis.

1.3.2. Transport of testosterone in the blood

As testosterone is lipophilic, it passes easily through membranes and leaves the Leydig cells by diffusion. In the blood, 98 % of testosterone is bound to transport proteins, and only 2 % is free and hence biologically active. Approximately 60 % of the circulating testosterone is bound with high affinity to the β-globulin sex hormone binding globulin (SHBG), and 38 % is loosely bound and transported by albumin. SHBG is a large glycoprotein (92.5 kDa) and is encoded by a gene on chromosome 17 in the immediate vicinity of the tumor-suppressant gene p53 (17p13.1). SHBG circulates in the serum as a homodimer, and carries two binding sites for sex steroids. Circulating SHBG is produced by the liver; however, SHBG is also locally produced by other tissues, e.g. the prostate and mammary gland[202]. SHBG shows a higher affinity for testosterone than for estradiol. Thus, increased production of SHBG by the liver causes a shift in the ratio of testosterone to estradiol by reducing the amount of free testosterone. As androgens reduce SHBG production, men have lower serum concentrations of SHBG than women. On the other hand, serum SHBG may be raised in men with testosterone deficiency. Other factors influencing SHBG production are listed in Table 1.1[323].

Stimulation of SHBG production	Inhibition of SHBG production
• Estrogen intake • Androgen deficiency • Growth hormone deficiency • Hyperthyroidism • Hepatitis • Liver cirrhosis • Phenytoin	• Androgen therapy • Overweight • Acromegaly • Hypothyroidism • Nephrotic syndrome • Corticosteroids • Hyperinsulinism • Gestagens

Table 1.1: Factors influencing sex hormone binding globulin (SHBG) serum concentrations.

Recent findings indicate that, in addition to transporting sex steroids, SHBG also has other functions. It may bind to specific, recently discovered, SHBG receptors on cell membranes, and hence possesses the ability to bind to testosterone or estradiol simultaneously. When a sex steroid binds to SHBG, the SHBG receptor complex is obviously activated by conformational changes, triggering an effect inside the cell. This may be the mechanism of the rapid non-genomic effects of testosterone and estradiol which occurs within minutes, and therefore may not follow the vival androgen and estrogen receptor-mediated pathways. It is unclear whether SHBG exerts autonomous effects equivalent to those of a hormone[157,324].

1.3.3. Metabolism of testosterone

Free testosterone diffuses passively into the target cells where it may be metabolized, depending on enzyme availability, to 5α-dihydrotestosterone (DHT) or 17β-estradiol (☞ Figure 1.5). In humans, 2 isoenzymes of 5α-reductase, which show large homology, are responsible for conversion to DHT. Type I 5α-reductase is encoded by a gene on chromosome 5. It is found mainly in the skin and liver. A gene on chromosome 2 carries the information for type II 5α-reductase, which is predominantly found in the prostate, adrenal gland, seminal vesicle, genital skin, hair follicles and cerebral cortex. Approximately 80 % of circulating DHT is produced by the peripheral conversion of testosterone, and 20 % is secreted directly by the testis.

> DHT and testosterone bind to the same androgen receptor, although DHT has a more than ten-fold greater affinity for the receptor and its dissociation from it is slower. Therefore, it is the more potent androgen. In physiological concentrations, the effects of testosterone and DHT complement one another. Therefore, both hormones are required for normal sexual development and virilization in puberty.

It is not clear whether DHT also induces effects which testosterone cannot, even in supraphysiological concentrations[99,136].

Furthermore, approximately 30 µg estradiol are produced per day by extratesticular aromatization of testosterone and androstenedione. Adipose tissue, bone cells and the prostate are rich in aromatase activity and can also produce estradiol.

In addition, approximately 10 µg estradiol are directly produced by the Leydig cells.

Hence, testosterone is the prohormone for DHT and estradiol.

> The complete spectrum of action of testosterone incorporates effects which are indirectly induced by conversion to DHT and estradiol (☞ Figure 1.6).

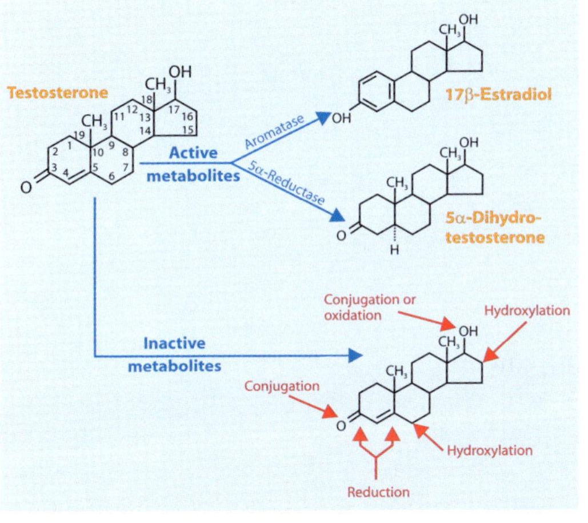

Figure 1.5: Metabolism of testosterone. In target organs, testosterone may be metabolized to 5α-dihydrotestosterone (DHT) or 17β-estradiol, depending on enzyme availability. Androgens are inactivated in the liver by oxidation, reduction or hydroxylation at different positions on the molecule.

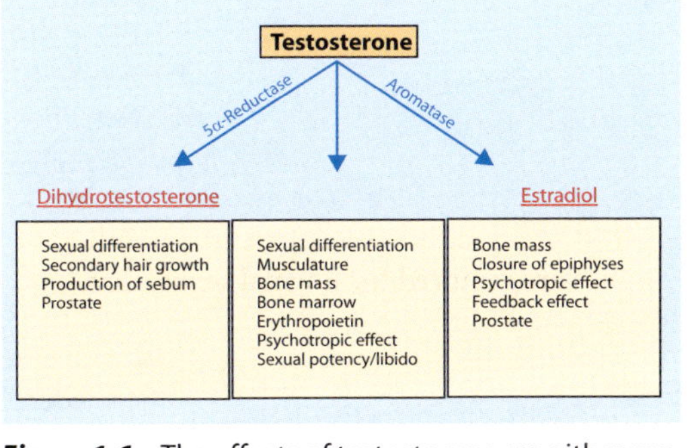

Figure 1.6: The effects of testosterone are either mediated directly (middle) or after metabolism to dihydrotestosterone (DHT) (left) or estradiol (right).

Testosterone and DHT are catabolized in the liver by two steps. First, the polarity of testosterone is increased by oxidation, reduction or hydroxylation. The initial and rate limiting step is reduction of the C4,5 double bond (☞ Figure 1.5). In the second step, the water solubility of the lipophilic steroid is facilitated by conjugation with glucuronic acid or sulfation on C3 or C17 of the androgen molecule. Elimination takes place via the urine as 17-ketosteroid and sulfate (e.g. androsterone, etiocholanolone, epiandrosterone, epitestosterone)[333]. Despite being bound to transport proteins in the blood, elimination of testosterone from the serum by the liver is very efficient - the half-life of free testosterone in the blood is only about 10 min.

1.3.4. Structure and function of the androgen receptor

The androgen receptor belongs to the family of steroid and thyroid hormone receptors, and is encoded by eight exons of a gene near the centromere on the long arm of the X chromosome (Xq11-12). The androgen receptor is a polypeptide of 910 amino acids with a molecular weight of 98.5 kDa[59].

Like the other steroid and thyroid hormone receptors, the androgen receptor is a DNA binding protein. Its activity is ligand-dependent and transcription-regulating. The androgen receptor carries three domains with different functions (☞ Figure 1.7).

The aminoterminal segment (exon A) inside steroid hormone receptors is very variable. It appears to affect transcription and to maintain the tertiary and quaternary structure of the androgen receptor.

The centrally located hydrophilic DNA-binding domains (exons B and C) carry two 'zinc' fingers, which bind to specific sections of DNA in or next to androgen-sensitive genes and thus influence transcription. The DNA-binding domains of the steroid hormone receptors are largely similar (40-90 %); the differences determine the gene specificity of the receptor.

The carboxyterminal end (exons D-H) carries the hydrophobic androgen-binding domain which is responsible for binding testosterone and DHT. This domain has a 40-50 % similarity with the amino acid sequence of the corresponding domains of gestagen, mineralocorticoid and corticosteroid receptors. However, there is only slight similarity with the ligand-binding domain of the estrogen receptor. Before binding with an androgen, the androgen receptor is associated with a heat shock protein (HSP 90) (☞ Figure 1.8).

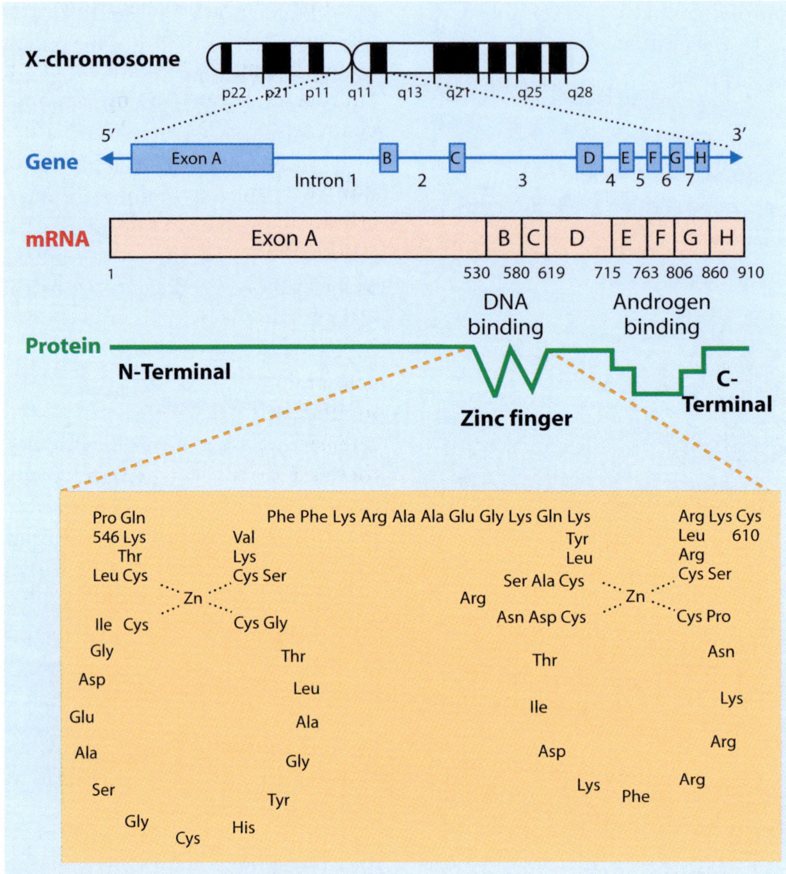

Figure 1.7: Location of the androgen receptor gene on the X chromosome in section Xq11-12. The gene is made up of eight exons (A-H) with seven introns in between. The mRNA encodes 910 amino acids. The DNA-binding domains (exons B and C) contain the zinc finger, which is presented in the magnified section in terms of its amino acid sequence. The zinc finger configuration is characteristic of all steroid hormone receptors. Exons D-H form the androgen-binding domain.

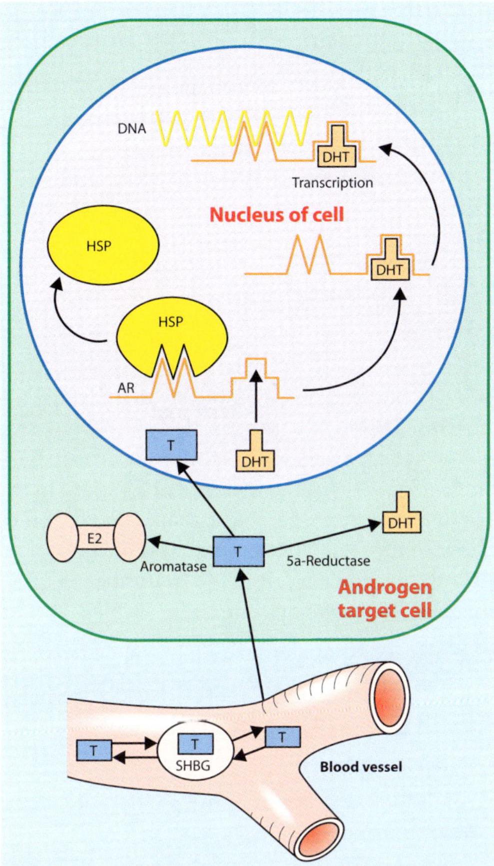

Figure 1.8: Intracellular mechanism of the androgen effect. Testosterone (T) diffuses into the cell and may be metabolized to 5α-dihydrotestosterone (DHT) or estradiol (E2), depending on enzyme availability. T or DHT binds to the androgen receptor (AR) associated with heat shock protein (HSP), which is dimerized after dissociation from HSP (not shown), migrates to the cell nucleus and binds to specific androgen-responsive elements of DNA by means of the zinc finger.

The significance of this association is unknown - HSP 90 may stabilize the androgen receptor or prevent its metabolism. Binding of testosterone or DHT to the androgen receptor induces a conformational change which causes dissociation of the androgen receptor from HSP 90 and activation of the androgen receptor. Activated androgen receptor complexes form dimers and then bind to specific DNA sections known as androgen-responsive elements (AREs), which lie in the promoter region of androgen-sensitive genes. Dimerization increases DNA-binding activity. The first zinc finger, in particular a group of three amino acids at its base, is responsible for specificity of the DNA binding[59]. The second zinc finger stabilizes androgen receptor binding. Binding of the androgen receptor to DNA influences the transcription of androgen-sensitive genes, which regularly lie downstream (3') of the ARE[313,240]. The transcriptional activity of the androgen receptor is modulated by numerous co-regulators[150].

Exon A contains a polymorphic polyglutamine (CAG) repeat sequence which normally has between 8 and 35 repeats. This sequence lies inside a split aminoterminal transactivation domain. The number of CAG repeats affects the transcription intensity of the receptor. The more CAG repeats there are the smaller is the activation of the transcription triggered by the androgen receptor. This is because co-activators and androgens bind more strongly to an androgen receptor with short CAG sequences than to androgen receptors with longer CAG repeats. Therefore, an androgen receptor with few CAG sequences mediates a stronger androgen effect than a receptor with many CAG repeats. A particularly large number of CAG repeats (> 38) is associated with mild androgen resistance, and Kennedy's syndrome, a degenerative bulbospinal motor neuropathy (spinocerebellar ataxia type 7)[448] (see Section 2.2.4.1).

As the prostate is an androgen-dependent organ, requiring testosterone and its metabolite DHT for growth and development, and the growth of prostate cancer is at least partially regulated by androgens, the association between the CAG repeat length in exon 1 of the androgen receptor and the incidence of prostate cancer has been studied extensively. However the results are still inconclusive. This might be due to very heterogeneous study populations concerning age, ethnic origin and phenotype of the prostate cancer[110,215,223,327,383,4402]. Even the studies finding an association between prostate cancer and the length oft the CAG repeats report only very weak associations[440] suggesting that the length of the CAG repeats as predictive parameter might not influence clinical practice in the near future. Larger studies addressing influencing factors like age, aggressivity of tumor growth, stage of disease, ethnic origin, and life style are needed to clarify the role of CAG repeats in the development and progression of prostate cancer.

Several cross-sectional studies on CAG repeat length and prostate size in eugonadal men have been done. The two most comprehensive studies compared matched healthy controls and BPH patients, and found an increased odds ratio for an enlarged prostate gland or BPH surgery in men with 19 or less CAG repeats [122,123]. Even though not all studies showed similar results, the development of BPH in eugonadal men is likely to be influenced by the length of the CAG repeats. Moreover prostate growth per year and absolute prostate size in hypogonadal men receiving testosterone therapy is strongly influenced by the CAG polymorphism of the androgen receptor with smaller treatment effects in longer repeats [446].

Furthermore there is an influence of CAG repeat length on flow-mediated vasodilatation and HDL levels in healthy volunteers, suggesting possible role of CAG polymorphisms of the androgen receptor in modulation of cardiovascular risk factors [445].

The data concerning the influence of the CAG polymorphism on bone metabolism and bone mineral density are inconsistent. Some studies report an influence of the CAG polymorphism others do not [208,445]. The available studies differ largely with respect to the used methods of measuring bone mineral density and the mean age of participants, possibly explaining the different results. However there seems to be a correlation between CAG polymorphism and bone density in younger men suggesting that CAG repeat might influence peak bone mass in men [211]. However, not all studies support this hypothesis [400].

Androgens are known to influence body fat mass. Two recent studies address the influence of CAG repeats on body composition. A German study on 106 healthy volunteers could demonstrate a positive association between the length of the CAG repeats and body fat. The longer the CAG repeats are the more fat mass the person has [447]. This was accompanied by an increase in insulin and leptin levels with increasing CAG repeat length. Furthermore fat free mass is influenced by the length of the CAG repeats as well. This has been demonstrated in two independent collectives [415].

Up to 80 % of men at the age of 80 years have developed male type baldness of different degree. In men with stronger baldness shorter CAG repeats have been described [331].

Taken together the CAG polymorphism of the androgen receptor modulates a variety of know androgen effects. But it remains unclear if CAG repeats will be a useful parameter to access individual risk for BPH or prostate cancer or other clinical conditions.

1.3.5. Biological effects of testosterone

Testosterone is not essential to survival, as is demonstrated by patients, completely deficient in the effects of testosterone as a result of an androgen receptor defect. However, testosterone is necessary for development of the male phenotype and in order to function properly as a man. In addition to classic androgen effects (sexual differentiation, virilization, potency), it exerts a number of other anabolic and metabolic effects (☞ Table 1.2)[263].

The effects of testosterone and its significance for development and function are age-related:

- In the embryo: sexual differentiation
- Puberty: virilization
- Adulthood: maintaining the phenotype, sexual function, anabolic effects

During the sexual differentiation phase, development and growth of the derivatives of the Wolffian duct, epididymis, vas deferens and seminal vesicle are promoted. In the adult male, testosterone, partially after conversion to 5α-dihydrotestosterone (DHT), serves to stimulate the epithelium and maintain the secretory function of these organs. Testosterone has an indirect effect on the prostate after conversion to DHT and estradiol by the enzymes 5α-reductase and aromatase, which can both be detected in prostate tissue. If no testosterone is produced, these organs start to regress, and the prostate appears conspicuously small on sonography. Failure of the seminal vesicle and prostate may lead to aspermia.

In the embryo, growth of the external genitalia, the penis in particular, is mainly dependent on the concentration of DHT produced *in situ* from testosterone. Lack of testosterone and lack of the enzyme 5α-reductase can lead to intersexual genitals or an abnormally small penis (micropenis). During puberty, the increase in androgens causes the

Target tissue	Effect	Active steroid
Wolffian duct	stimulates growth and differentiation	Testosterone
External genitalia	masculinization and growth	DHT
Urogenital sinus	masculinization and growth	DHT
Bones	closure of epiphyses, anabolic effect	estradiol and testosterone
Larynx	growth with lengthening of vocal cords	testosterone or DHT
Skin	stimulates sebum production stimulates growth of body hair and facial hair reduces main hair growth (androgenetic alopecia)	DHT DHT DHT
Kidneys	stimulates erythropoietin production	testosterone/DHT
Liver	induces enzymes, influences protein synthesis	testosterone or DHT
Bone marrow	stimulates erythropoiesis	testosterone/DHT
Musculature	anabolic effect	testosterone
Testes	stimulates and maintains spermatogenesis	DHT/estradiol
Prostate	stimulates growth and function	DHT/estradiol
Breasts	inhibits growth	testosterone or DHT
Hypophysis	negative feedback to gonadotropin secretion	testosterone/DHT
Hypothalamus	negative feedback to GnRH secretion	DHT
Brain	psychotropic effects including on libido	testosterone/DHT/estradiol

Table 1.2: Androgen target tissues and androgen effects. These effects are either directly mediated through testosterone or after conversion to 5α-dihydrotestosterone (DHT) or estradiol, GnRH, gonadotropin releasing hormone.

penis to grow in size. It is not possible to stimulate further penile growth after puberty as there are no androgen receptors present after this time. On the other hand, androgen deficiency in the adult does not cause the penis to shrink.

Androgens have a very specific and differentiated effect on the growth of body hair. Armpits and the pubic area contain high levels of the enzyme 5α-reductase and therefore produce a large amount of DHT, which stimulates the growth of hair follicles in these areas. The skin on the face contains less 5α-reductase, and therefore facial hair growth is stimulated by high concentrations of testosterone. Therefore facial hair appears at a later stage of puberty than pubic and armpit hair. However, growth of the hair on the head is inhbited by androgens. Hence, the typical frontotemporal (temple) hair loss experienced in males is a physiological effect of testosterone (after conversion to DHT). This is modulated by genetic factors, local 5α-reductase activity and the presence of androgen receptors. Therefore, hair loss in men means that sufficient androgen is present for androgenization. Men who have been suffering from hypogonadism for some time do not present with baldness or loss of hair at the temples.

Sebaceous glands in the skin are also stimulated. Thus. acne may develop for a short while during puberty and at the start of testosterone substitution. In contrast, androgen deficiency causes the skin to become dry, thin and sensitive[235].

During puberty, testosterone stimulates the larynx to grow and the vocal cords to elongate, which causes the voice to drop or 'break'. Once the vocal cords have stretched, this process is irreversible, and therefore androgen deficiency in an adult does not cause the voice to become high-pitched again. However, if necessary, in men with late discovered hypogonadism, even in the third and fourth decade of life, the pitch of voice can be lowered.

Muscle is a well-known target organ of testosterone. Testosterone stimulates mRNA and protein synthesis and leads to an increase in muscle mass through the growth of muscle fibers. There is no evidence to suggest that androgens will cause new muscle fibers to form. Increased muscle mass also means increased muscle strength. Anabolic effects appear to be induced exclusively by testosterone, but not by 5α-dihydrotestosterone, as men with type II 5α-reductase deficiency appear to have normal muscle mass. Also, administration of 5α-reductase inhibitors (e.g. finasteride) does not cause muscle atrophy[41,399].

For the bone metabolism, the action of testosterone appears to be of less importance than its metabolite estradiol. Estrogens are critical to mineralization and hence bone density; these are produced *in situ* in the bones from testosterone. Osteoblasts contain the enzyme aromatase and can therefore convert androgens to estrogens[382]. This is of enormous significance to bone density, as evidenced by the discovery of a man with estrogen receptor deficiency and men with genetically induced deficiency of the enzyme aromatase[71,365]. Despite high concentrations of testosterone, men with either disease suffer from severe osteoporosis, obviously due to the absence of estrogen action. Estrogens also appear to be responsible for sealing the epiphyses during puberty. Men in whom the estrogen effect is absent have open epiphyses and a high growth rate, typical of eunuchs. Administration of estrogens in aromatase deficiency closes the epiphyses and causes height growth to cease. Therefore, testosterone is the prohormone in bones for conversion to estrogen, which then exerts anabolic effects. In addition, testosterone itself exerts some effects, although these are less significant: it inhibits the recruitment of osteoclasts and stimulates osteoblasts which carry androgen receptors[82,294].

Testosterone promotes hematopoiesis in two ways: it stimulates both renal and extrarenal production of erythropoietin. It also directly promotes hematopoiesis by its specific action on bone marrow. More stem cells and erythropoietic precursor cells are recruited and programmed to differentiate to red blood cells. The effect appears to be triggered by testosterone and 5α-dihydrotestosterone in equal measure. Together these effects cause more red cells to form. This results in increased hematocrit and hemoglobin levels in the blood, which is why the reference ranges for these parameters are higher in men than in women.

In the liver, the activity of numerous enzymes is affected by androgens. Hepatocytes carry androgen receptors and are therefore a direct target organ of testosterone and DHT[102]. Hepatic microsomal enzymes involved in the breakdown of many drugs are generally stimulated by androgens. This may mean that certain established medications need adjustment (e.g. antiepileptics or coumarin). In contrast, the activity of steroid metabolism enzymes such as 5α-reductase and hydroxylases is reduced. Testosterone also decreases hepatic production of specific binding proteins for steroids and thyroid hormones (SHBG, cortisol binding globulin (CBG) and thyroxin binding globulin (TBG))[263].

The effect of androgens on lipid metabolism has attracted a lot of attention. It might at least partially explain the higher incidence of atherosclerosis, and the shorter lifespan of men compared with women.

Extensive epidemiological studies by the Lipid Research Clinic of North America show that, until puberty, there are no gender-specific differences in high-density lipoprotein cholesterol (HDL cholesterol), low-density lipoprotein cholesterol (LDL cholesterol) and triglyceride levels in children. During puberty, HDL cholesterol falls in boys, and LDL cholesterol and triglycerides go up[274,275]. Not until men reach their 50s or 60s does HDL cholesterol increase, possibly as a result of the age-related reduction in testosterone. Testosterone substitution in men with hypogonadism also causes HDL cholesterol to decrease and LDL cholesterol to increase, and therefore the risk of atherosclerosis increases. Furthermore, there is evidence, yet to be confirmed, that androgens increase fibrinolysis by stimulating plasminogen and reducing plasminogen activator inhibitor[74,362]. This may reduce the adverse effects on lipid metabolism. An enormous amount of research still needs to be carried out in this area to clarify the role of androgens on atherosclerotic risk factors. In addition, their effects on lipoprotein (a) are unclear. Furthermore, in aging men, the substitution of testosterone leads to opposite effects as in young men. Total cholesterol as well as LDL cholesterol decrease when testosterone is applicated. Currently, it is unclear why testosterone has positive effects on lipid metabolism in elderly men, divergent from its effects in young men.

The mechanism behind the androgen-induced effect on lipid metabolism is still partially unclear, although it has been demonstrated that androgens increase the activity of hepatic triglyceride lipase, hence accelerating the breakdown of HDL cholesterol[263]. However, it has not yet been shown whether androgens also affect hepatic expression of the LDL cholesterol receptor. This effect may be mediated by the androgen receptors identified in hepatocytes[102].

Androgen receptors are found in the hypothalamus, hypophysis, cerebellar tonsil and septum of the central nervous system. In addition, the enzymes aromatase and 5α-reductase are found in nearly all regions of the brain, although to varying extents. Therefore, it is difficult to distinguish between the effects of testosterone, DHT and estradiol. In the hypothalamus and hypophysis, androgens exert a negative feedback effect on GnRH, LH and FSH secretion, thus closing the dynamic feedback loop between the testes and the higher centers of the brain. Whereas it has been known for some time that animals possess a morphologically recognizable sexual dimorphism of the brain with varying volume of brain centers, this has not yet been clearly shown in humans. Nevertheless, androgens are known to exhibit a number of psychotropic effects: not only are general well-being and physical and mental ability influenced by androgens, but drive and emotional state are too[158]. A lack of testosterone causes depression and lack of drive, and restricts mental and physical ability. There have been isolated reports of men on high-dose anabolic steroids attempting to commit suicide after halting them abruptly. The extent to which other behaviors and cognitive abilities such as aggression, spatial and mathematical cognition and verbal competence are influenced by androgens is currently being researched. There is evidence that androgens increase aggressive behavior and spatial cognition.

On the other hand, the effect of androgens on sexual behavior and activity is well-documented. Sexual appetite, fantasies, spontaneous nightly erections and sexual activity as well as frequency of orgasm and ejaculation are also supported by androgens. However, erectile potency in the context of sexual activity clearly appears to be less dependent on testosterone, which is why testosterone therapy has hardly any effect on erection if serum testosterone levels are normal or only slightly reduced.

1.3.6. Testosterone and the metabolic syndrome

Currently there is no unique definition of the metabolic syndrome[50]. The *interheart* study, one of the most comprehensive case control studies for identifying risk factors for myocardial infarction highlighted abdominal obesity as a major risk factor for cardiovascular events[439]. According to the national cholesterol education board in men a waist circumference of 102 cm or more, and 88 cm in women is a main criterium for defining the metabolic syndrome[140]. Furthermore, two out ouf the following criteria have to be present for establishing the diagnosis of the metabolic syndrome: blood glucose >5.6 mmol/L (>100 mg/dl), HDL-cholesterol <1.0 mmol/L (men), triglycerides >1.7 mmol/L, blood pressure >130/85 mmHg. In 2004 the International Diabetes Federation redefined the metabolic syndrome and used ethnicity specific cut off values for the waist circumference, meaning for Caucasian men and women upper limits of 94 cm, and 80 cm, respectively[161].

Since abdominal obesity is closely correlated to insulin resistance, the American Association of Clinical Endocrinologists and the "Group for the Study of insulin resistance" highlighted the role of insulin resistance as underlying pathophysiological basis for the metabolic syndrome[21]. Recently, this concept came into question, since only 48 % of patients with insulin resistance developed the metabolic syndrome[50,254]. However there is no doubt that abdominal obesity is a major risk factor for hypertension, diabetes mellitus and myocardial infarction.

Recent molecular biological research identified the adipose tissue as an endocrine active organ. One of these hormones, leptin, is produced by adipocytes and binds to specific hypothalamic receptors, suppressing appetite and regulating energy expenditure[442]. Thus, mice lacking functional leptin receptors develop obesity and glucose intolerance. Interestingly, not only obesity but also infertility occurs in mice which are unable to mediate leptin signals in the hypothalamus. Brain-/neuron-specific insulin receptor knockout (NIRKO) mice which show due to genetic manipulation insulin resistancy specifically in the brain develop obesity and reduced fertility. According to these mouse models energy metabolism is tightly connected to reproductive function, suggesting that also in humans obesity and insulin resistance might influence sex hormone levels.

In the last years several epidemiological studies have been performed to elucidate the correlation between obesity, insulin resistance, diabetes mellitus and testosterone levels in men. One of the first and largest trials was the Telecom Study on 1292 healthy men showing an inverse relationship between total testosterone and insulin levels. These data suggest that high insulin levels or a possible underlying insulin resistance might influence testosterone concentration in healthy men or vice versa[356]. These results were reproducible in other study groups[289]. Furthermore, there are numerous studies which demonstrate an association between fasting glucose levels, diabetes mellitus and low total and free testosterone levels [23,142,372,390,443], suggesting that an impaired glucose tolerance could also influence testosterone levels. However, several studies show an association of abdominal obesity – the most common cause of insulin resistance – and low testosterone levels. One of the largest trials was the Tromsø study which investigated 1548 men aged 25–84 years[377]. In age- and body mass index(BMI)-adjusted analyses, men with a waist circumference ≥102 cm had significantly lower levels of total testosterone and SHBG compared to men with an ideal waist circumference, defined as <94 cm. The adjustment to BMI in this study diminished, but did not eliminate the association between total/free testosterone levels and waist circumference. The lowest levels of total and free testosterone were observed in men with relatively high waist circumference despite relatively low overall obesity (BMI), suggesting that waist circumference is the higher-ranking parameter for prediction of endogenous testosterone levels[377].

Another line of evidence concerning a central role of testosterone in regulating important metabolic processes was studied in men receiving androgen deprivation therapy for the treatment of prostate cancer[366]. Smith et al. investigated 22 men treated with GnRH analoga for prostate cancer. Only one month after initiation of therapy serum fasting insulin levels increased significantly. The increase of insulin levels correlated with an increase in body fat mass. A further short-term study over 12 weeks on 25 men receiving GnRH agonists showed that androgen deprivation leads to an increase in body fat mass and a decrease in insulin sensitivity using the HOMA IR model[367]. In a larger group under long-term surveillance for up to 5 years the increase of body fat mass during androgen withdrawal was confirmed [77]. In summary androgen deprivation in men with prostate cancer leads to an increase in body fat mass and insulin resistance. That this is a consequence of androgen deprivation rather than of prostate cancer itself was demonstrated clearly by Basaria et al. who had two control groups: one group of healthy men and one group of prostate cancer patients who received treatment other than androgen deprivation. Increases of fasting insulin, fasting glucose, and insulin resistance occurred only in men on androgen deprivation therapy[25]. A further line of evidence for a role of testosterone in the development of insulin resistance is added by the study of Pitteloud et al.[307]. Their detailed investigation of 60 men with normal glucose tolerance, impaired glucose tolerance and with diabetes mellitus type 2 demonstrates a positive correlation between serum testosterone levels and insulin sensitivity in men across the full spectrum of glucose tolerance. Moreover, men with hypogonadal testosterone levels were twice as insulin resistant as their eugonadal counterparts, and 90 % fulfilled criteria for the metabolic syndrome. From a clinical perspective, these data highlight the importance of performing a comprehensive metabolic evaluation in men with hypogonadism. In addition, the data showed a correlation of testosterone levels with genetic (OXPHOS gene expression) and functional markers of mitochondrial function suggesting a novel molecular mechanism whereby testosterone might modulate insulin sensitivity in men[307].

Testosterone treatment in male hypogonadism is recommended to maintain libido, erectile function and to prevent negative effects of long-term androgen deficiency, i. e. muscular atrophy, osteoporosis and anemia. Treatment of hypogonadal men results in a decrease of total body fat. One of the longest trials was performed by Wang and coworkers[418]. They investigated 163 hypogonadal men treated with testosterone gel 5, 7.5 or 10 g/d. 70 % of the subjects were using this medication for at least 30 months in total. Fat mass decreased by -0.8 ± 0.3 kg after 6 months and -1.57 ± 0.38 and -1.30 ± 0.51 kg at 18 and 30 months of treatment, respectively. Several studies investigated the effect of testosterone on body composition in aging males. The beneficial effect of testosterone treatment on the reduction of body fat mass is also present in older men. A recent clinical study with old men (mean age 71 years) and low testosterone levels clearly demonstrated that treatment with testosterone leads to a decrease of fat mass by about 2 kg, which was maintained throughout the entire study period of 3 years[295]. A recent meta-analysis of these studies concluded that testosterone treatment leads to decrease of about 1.5 kg in aging males[166]. Furthermore, this improvement of body composition was not only accompanied by improved physical performance but also by significant decreases of total cholesterol, LDL-cholesterol and triglycerides. HDL-cholesterol remained unchanged, as did fasting insulin and adiponectin levels[295]. The positive effect on total and LDL-cholesterol of testosterone treatment in older men or in men with abdominal obesity had also been shown previously[241,243,244,441]. The effect of testosterone is dose dependent. In 61 young eugonadal men, randomized to one of five groups to receive monthly injections of a long-acting GnRH agonist, to suppress endogenous testosterone secretion, and weekly injections of 25, 50, 125, 300 or 600 mg of testosterone enanthate for 20 weeks the changes in fat-free mass were highly dependent on testosterone dose and correlated with testosterone concentrations[44].

In summary, it is well accepted that testosterone plays an important role in the metabolic syndrome. The presence of metabolic syndrome is frequently associated with low testosterone levels. However, it still remains controversial whether low testosterone levels are a consequence of the metabolic syndrome or a causative factor. Recent evidence as well as epidemiological studies suggest that testosterone might influence insulin sensitivity beyond its effect on body composition. Six large prospective studies have shown that low testosterone levels predict development of type 2 diabetes in men[142,225,289,372,377,390]. Two small short term studies suggested that testosterone therapy in men with obesity or diabetes mellitus type 2 was associated with an increase in insulin sensitivity[243], although a larger clinical study did not confirm this observation[359]. Although there is no clear cut evidence, the present information points towards testosterone being a modulator of insulin sensitivity with low levels of testosterone increasing the likelihood of insulin resistance and a beneficial effect of testosterone treatment in men with metabolic syndrome and low testosterone levels.

1.4. From zygote to man

One of the most interesting questions relating to ontogenesis is the clarification of the mechanisms behind sexual differentiation. How does man become a man? This can be answered in regard to the establishment of the gender in an embryo. The application of biomolecular technology over the past decade, based on work carried out in the 1960s and 1970s, has produced conclusive findings on the genetic control of gonadal differentiation and the effect of hormones on development of the phenotype. This has enabled numerous clinical syndromes which were formerly grouped under the collective term of male pseudohermaphroditism to be explained and classified.

1.4.1. Sexual differentiation

Sexual differentiation is based on a complicated, sequential interaction of genetic and hormonal influences on the embryo. The structures of the gonads, ducts and external genitalia are identical in both sexes until the 6th week of gestation and may develop into either male or female. The biological 'default' of all three structures is development into the female phenotype. Additional genetic and hormonal factors are involved to enable differentiation to the male phenotype. If these distinguishing factors are absent, or the target tissue does not react adequately, the tissue in question will automatically become feminized.

Gender is evaluated according to three criteria:

- Chromosomal (genetic) gender
- Gonadal gender
- Somatic gender (phenotype)

Chromosomal gender is laid down when the gametes fuse. In humans, as in most mammals, the homogametic gender is female (XX) and heterogametic gender male (XY). The Y chromosome induces differentiation of the gonadal structure to testes, which determines gonadal gender. The hormones of the testes then determine somatic gender by inducing differentiation to the male phenotype.

1.4.1.1. Differentiation of gonadal structure

As soon as an intact Y chromosome is present in the karyotype, the gonadal structure differentiates to testes regardless of the number of X chromosomes. A gene on the distal section of the short arm of the Y chromosome is responsible for this (the so-called 'sex-determining region Y' (SRY). SRY encodes a protein that is 204 amino acids in length. Among other things, this carries a motif which is 79 amino acids in length (high-mobility group, HMG box), and is attributed with DNA-binding and DNA-bending properties. SRY appears to regulate the transcription of autosomal genes by binding to DNA and bending it such that normally distant sections of DNA lie adjacent to one another, and therefore other transcription factors can dock onto the newly created binding site. Interacting transcription factors include the gene products of DAX-1, SOX9, SF1 and WT1. The precise mechanisms are still unknown, as are the autosomal genes downstream[69,238]. One intensively investigated candidate is the WT-1 gene (Wilms tumor gene). Mice with a null mutation for WT-1 do not develop testes despite an intact SRY. Denys-Drash syndrome is characterized by kidney disease in early childhood, the development of a Wilms tumor and gonadal dysgenesis (defective gonadal development). It results from a mutation of the WT-1 gene[253].

According to morphological criteria, it is not possible to distinguish later male and female gonads in an embryo until approximately the 42nd day of gestation. Presumably, SRY induces differentiation of the Sertoli cells in the 7th week of gestation; these, together with primordial gametes, then form tubular structures, precursors to the seminiferous tubules. Leydig cells can be observed in the testis of the embryo only towards the end of the 8th week of gestation, and testosterone biosynthesis starts in the 9th week of gestation. The prerequisite for differentiation of somatic gender to the male phenotype is satisfied with differentiation of the endocrine-active Sertoli and Leydig cells. This is exclusively controlled by hormones in the testis. The ovary plays no active role in the development of somatic gender.

1.4.1.2. Differentiation of somatic gender

The shape of the internal and external genitalia characterizes somatic gender. Three hormones are required to differentiate the ducts which have the same structure in both sexes (Wolffian duct and Müllerian duct), and the urogenital sinus to a male phenotype: anti-Müllerian hormone (AMH, previously known as Müllerian-inhibiting factor (MIF), testosterone and 5α-dihydrotestosterone (DHT). AMH and testosterone control differentiation of the ducts. DHT, together with testosterone, controls the shape of the urogenital sinus (☞ Figure 1.9).

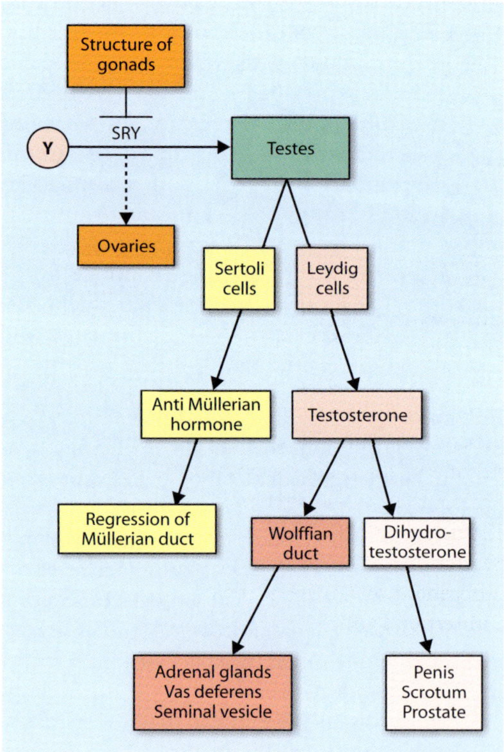

Figure 1.9: Diagram of sexual differentiation. The 'sex-determining region Y' (SRY) of the Y chromosome induces differentiation of the gonads to testes, determining gonadal gender. The testicular hormones anti-Müllerian hormone (AMH), testosterone and 5α-dihydrotestosterone (DHT) then establish somatic gender by inducing atrophy of the Müllerian duct and differentiation of the Wolffian duct, urogenital sinus and external genitalia to the male phenotype.

Until the 7th week of gestation, the embryo displays both the Müllerian duct and the Wolffian duct. The uterus and Fallopian tubes form from the Müllerian duct, while the Wolffian duct forms the structure for the epididymis, vas deferens and seminal vesicle. Differentiation of the Müllerian duct to uterus and Fallopian tubes occurs automatically even in the absence of ovaries or testes. However, if testes have been formed, the Müllerian duct atrophies by the action of AMH. AMH is a glycoprotein formed from two identical subunits bound by disulfide bridges. It is structurally similar to inhibin and transforming growth factor β. The gene for AMH lies on the short arm of chromosome 19. AMH reaches the Müllerian duct via an endocrine and a paracrine route, and induces its complete regression within a few weeks (by means as yet unknown). If AMH is absent, the Müllerian duct will turn into the uterus and Fallopian tubes, regardless of genetic and gonadal gender. Androgens have no influence on the Müllerian duct.

In contrast with the negative regulation of the Müllerian duct (inhibition by AMH), the Wolffian duct is subject to positive regulation by testosterone. Under the influence of hCG from the mother, testosterone biosynthesis in the Leydig cells starts in the 9th week of gestation. Testosterone blood levels reach a maximum in the 9th-14th week of pregnancy and fall to the same level as in female fetuses after the 24th week (☞ Figure 1.10). As hepatic production of SHBG is low in the embryo, the concentration of free, biologically active testosterone in the male embryo at the time of sexual differentiation exceeds that of the adult male. Testosterone enters into the Wolffian duct via an endocrine and a paracrine route, and induces its differentiation to the adrenal gland, vas deferens and seminal vesicle.

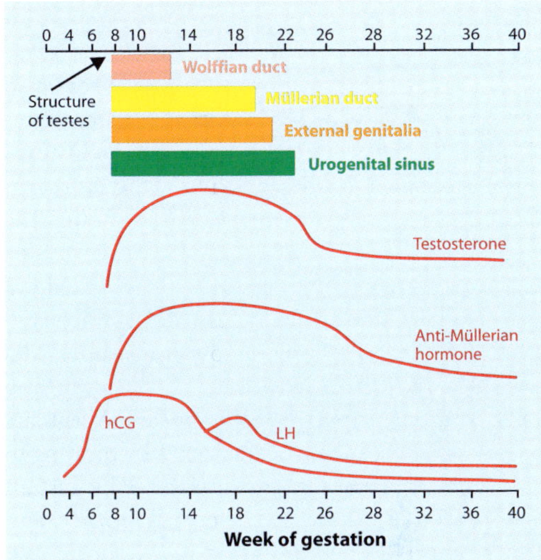

Figure 1.10: Time-line of differentiation of ducts and sexual organs with testicular production of anti-Müllerian hormone (AMH) and testosterone in male fetuses.

1.4.1.3. Differentiation of external genitalia and urogenital sinus

Until the 8th week of gestation, the external genitalia of both sexes are identical and may become either male or female. The genital protruberance (structure of the clitoris and the glans penis with corpus cavernosum) is surrounded at its attachment by the genital labia (structure of labia majora and scrotum) (☞ Figure 1.11).

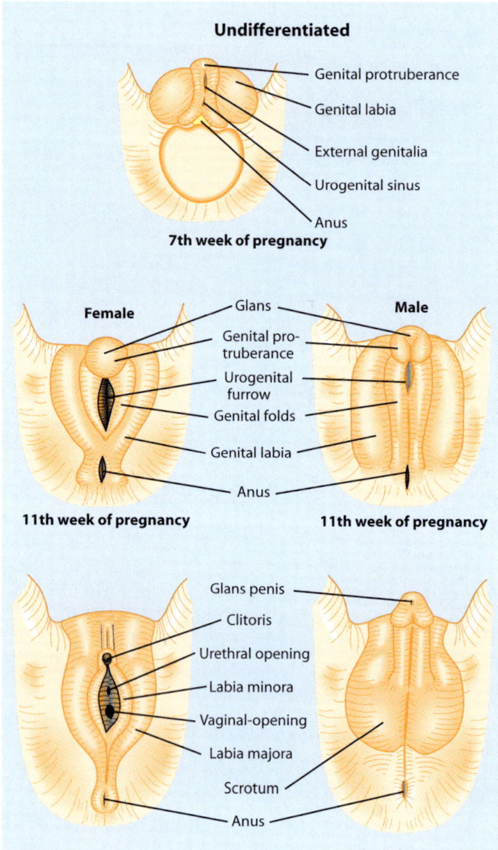

Figure 1.11: Differentiation of the external genitalia of male and female fetuses from initially identical structures in both sexes.

The underside of the genital protruberance shows pairs of genital folds (structure of labia minora and corpus spongiosum) which surround the urogenital ridge. Under the influence of DHT, the bipotent structure becomes differentiated to the male phenotype. Testosterone transported in the blood to the target organs functions as a prohormone and is converted to DHT by the enzyme 5α-reductase, which is highly active in the structure of the external genitalia. DHT induces growth of the genital protruberance, and fusion of the genital folds and genital labia. At the same time, DHT causes differentiation of the prostate, which arises out of the urogenital sinus. Differentiation is complete by approximately the 14th week of gestation. However, the external genitalia, especially the penis, which is the same size as the clitoris in the 14th week of gestation, continue to grow under the influence of DHT. If the DHT effect is absent, for example either due to absent testis or testosterone production being disturbed, testosterone not metabolizing to DHT (5α-reductase deficiency) or defective androgen receptors, masculinization will be incomplete. The degree will depend on the extent to which the DHT effect is absent. The clinical picture of a disorder such as this ranges from discrete hypospadia to complete phenotypical feminization.

Production of testosterone in the fetus reaches a peak at the time of differentiation of the external genital organs and urogenital sinus (between weeks 9 and 14 of gestation). It then falls again (☞ Figure 1.10)[106]. In the second half of pregnancy, LH from the hypophysis of the fetus takes over stimulation of the Leydig cells[120]. If stimulation is insufficient, e.g. as a result of secondary hypogonadism (idiopathic hypogonadotropic hypogonadism, Kallmann's syndrome), the genitalia do not grow, resulting in an abnormally small penis.

1.4.2. Development during puberty

After birth, further sexual development and differentiation remain dormant throughout childhood. The mechanisms triggering the start of puberty are unclear. Leptin, a hormone produced by adipocytes tissue, may be a trigger, in that it indicates adequate body mass for initiation of reproduction via a positive feedback to the hypothalamus. Puberty begins with the pulsatile secretion of GnRH by the hypothalamus (initially only at night). This then triggers the production and release of the gonadotropins LH and FSH. Hence, peaks of LH and later FSH can be detected sporadically at night only. LH enables the slowly increasing production of testosterone in the testis. Thus, the first sign of the start of puberty is growth of the testes and scrotum, whereby the age at which boys reach stage G2 (testis volume > 3 ml) (☞ Table 1.3) varies enormously (☞ Figure 1.12). 90 % of all boys reach stage G2 between the ages of 9 and 13. Shortly afterwards, the first pubic hair can be seen at the base of the penis.

Genital development	
Stage	Description
G1	child-like genitals, smooth, pale scrotal skin, small penis, volume of testis less than 3 ml
G2	scrotum and testis grown a little, scrotal skin curly
G3	penis grows, first in length, then in circumference further growth of testis and scrotum
G4	development of glans penis, scrotal skin starting to become pigmented and more curly
G5	Adult genitals in terms of size and structure
Growth of pubic hair	
Stage	Description
P1	pre-pubic, no pubic hair
P2	single long, slightly pigmented hair at base of penis which is not curly
P3	darker, curlier hair on mons pubis
P4	adult hair on mons pubis but still no hair on inner thighs
P5	adult hair in area of mons pubis, also starting to grow on inner thighs but still not extended to the abdomen (= horizontal line of pubic hair)
P6	extension of pubic hair to abdomen in diamond pattern

Table 1.3: Standard stages of puberty in boys (☞ Figure 1.13)[227].

Each stage of puberty can last for a varying amount of time. On average, puberty (stages G3-G5) lasts 3.5 ± 1.1 years (mean ± standard deviation), and stages P2-P5 2.7 ± 1.0 years. However, almost one-third of all men have not reached stage P6 in growth of pubic hair by 20 years of age. Nevertheless, growth of the testes can vary enormously (☞ Table 1.4). The pubertal growth spurt in height regularly occurrs in the last third of puberty.

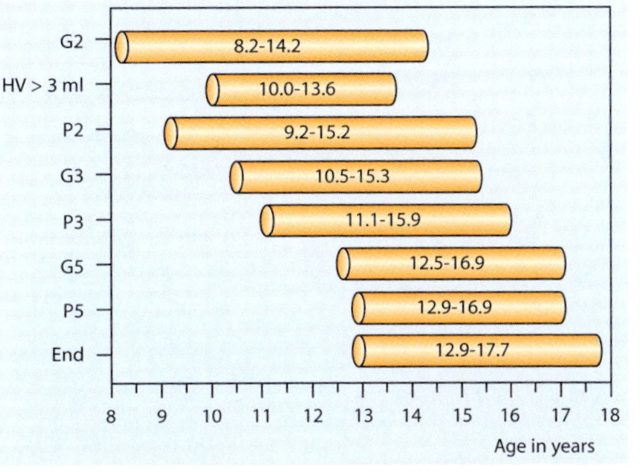

Figure 1.12: Age range at which healthy boys go through stages of puberty[227]. Data are based on studies in 142 Swiss boys and indicate a range of 2 standard deviations. G2, stage G2 genital development; G3, stage G3; G5, stage G5; TV, volume of testes; P2, growth of pubic hair stage P2; P3, stage P3; P5, stage P5; end, testes stop growing.

Age (years)	Median (ml)	Mean ± standard deviation (ml)	Volume (ml)
10	1.3	1.4 ± 0.5	1-3
11	2.0	2.1 ± 0.9	1-6
12	3.1	3.7 ± 2.1	1-12
13	4.9	5.8 ± 3.5	1-20
14	8.4	9.2 ± 4.5	1- ≥ 25
15	11.9	12.5 ± 4.6	2- ≥ 25
16	14.7	14.5 ± 4.7	4- ≥ 25
17	19.6	18.2 ± 4.7	6- ≥ 25
19	19.9	19.5 ± 4.7	8- ≥ 25

Table 1.4: Testicular volume (ml) during puberty (right testicle)[227].

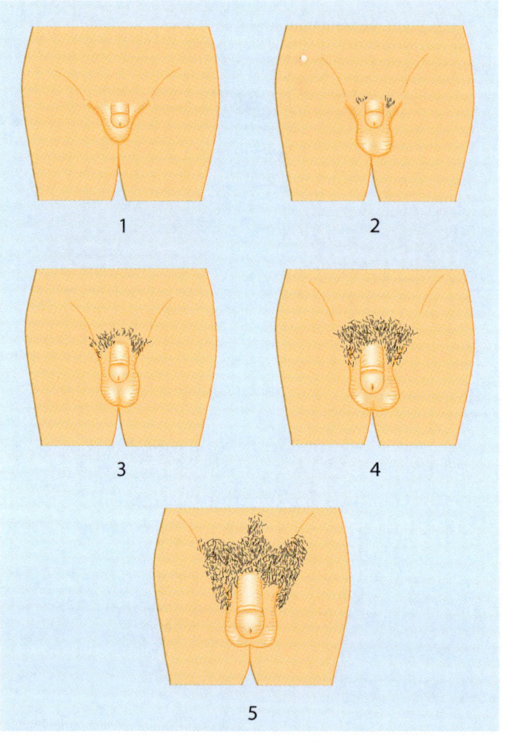

Figure 1.13: Diagram of stages of puberty in boys according to Tanner. The numbers refer to the corresponding stages in genital and pubic hair growth as described in Table 1.3.

Hypogonadism in men

2. Hypogonadism in men

Hypogonadism is inadequate function of the testes in man. The testes perform a number of functions in the development and maintenance of male characteristics and reproduction. These include:

- The production of endocrine and paracrine hormones (e.g. testosterone, Müllerian inhibiting hormone) which are necessary in utero for sexual differentiation;
- The production of mature sperm (spermatogenesis) which are required to fertilize an egg cell;
- the synthesis and secretion of steroid hormones (androgens) which are needed for sexual maturity, virilization, the expression of male behavior as well as anabolic effects;
- The production of other, non-steroidal hormones (inhibin, activin) which influence the dynamic feedback control system to the hypothalamus and hypophysis.

If the cause of hypogonadism lies in the testis itself, this is referred to as primary or hypergonadotropic hypogonadism. Hypothalamic or pituitary disorders of testicular function are grouped together under the term secondary or hypogonadotropic hypogonadism (☞ Figure 2.1).

2.1. Clinical picture of hypogonadism

The clinical picture of androgen deficiency is dependent on the time at which this deficiency appears and the extent of the deficiency (☞ Table 2.1).

In the fetus

Androgen deficiency or the absence of an effect from androgens during sexual differentiation (9th-14th week of gestation) leads to intersexuality with insufficient or absent masculinization of the external genitalia. The extent of intersexuality ranges from a completely female phenotype to external male genitalia with distal hypospadia. Insufficient effect from testosterone in the later phase of development (up until about the 24th week of pregnancy) can cause abnormal positioning of the testes and an abnormally small penis (micropenis).

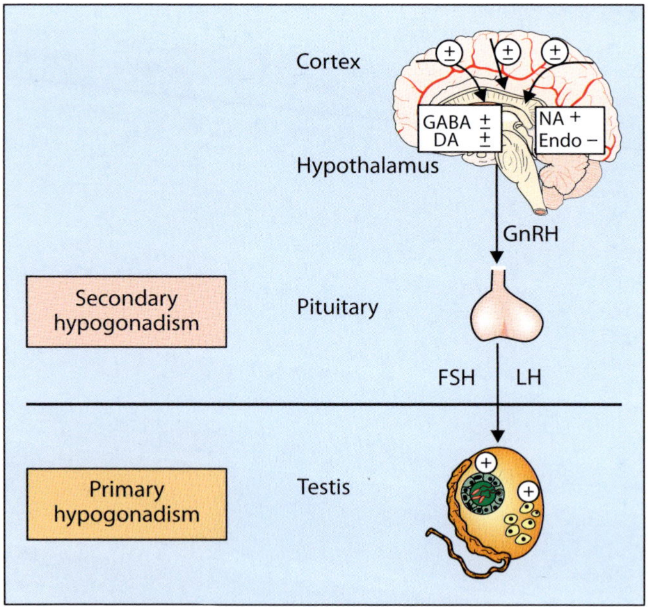

Figure 2.1: Distinguishing between primary (testicular) and secondary (hypothalamic or pituitary) hypogonadism.

Organ	Before end of puberty	After end of puberty
Bones	excessive eunuchoid growth osteoporosis	osteoporosis
Larynx	voice does not break	no change in voice
Hair	feminine hair type: • horizontal pubic hair line • hair at temples straight across • no facial hair • no body hair	decreasing facial, armpit, pubic and body hair, no androgenic alopecia (at temples)
Skin*	dry skin and no acne in puberty as no sebum produced, pallor	no sebum produced, atrophy, pallor, fine wrinkles
Erythropoiesis	anemia	anemia
Musculature	underdeveloped, no strength	atrophy, decreasing strength
Fat distribution	female (pronounced hips)	increasingly female
Lipid metabolism	HDL-C $\uparrow$, LDL-C $\downarrow$	HDL-C $\uparrow$, LDL-C $\downarrow$
Spermatogenesis	not initiated, infertility	stops, increasing infertility
Semen	usually aspermia	small volume
Libido and potency	do not develop	reduced or absent
Penis*	child-like	no change in size
Scrotum*	not very pigmented, slightly wrinkled	no change
Prostate*	small, underdeveloped	atrophy

Table 2.1: Signs of androgen deficiency[284].
*The androgenic effect of 5α-dihydrotestosterone is exerted in these tissues (formed from testosterone in the target tissue); HDL-C, high-density lipoprotein cholesterol; LDL-C, low-density lipoprotein cholesterol.

Puberty

If testosterone deficiency occurs after birth but before puberty, virilization will not occur. This results in the typical syndrome of eunuchoidism. Epiphyseal cartilage does not stop growing, so the long bones of the arms and legs continue to grow. The long bones of the legs grow at a quicker pace than the trunk (ratio of upper to lower body in cm < 1, gigantism), and the arms extend beyond the torso by over 5 cm (☞ Figure 2.2).

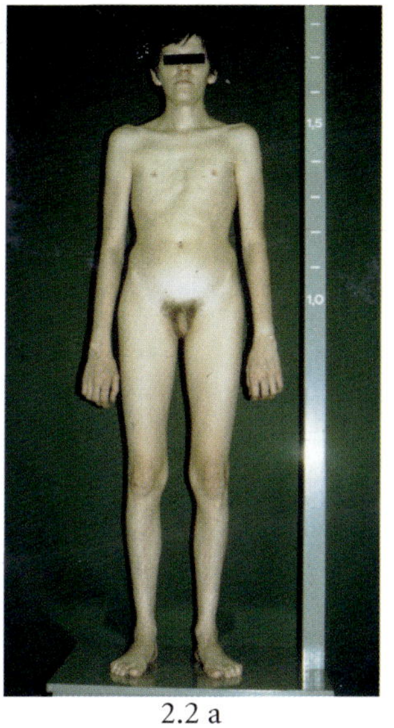

2.2 a

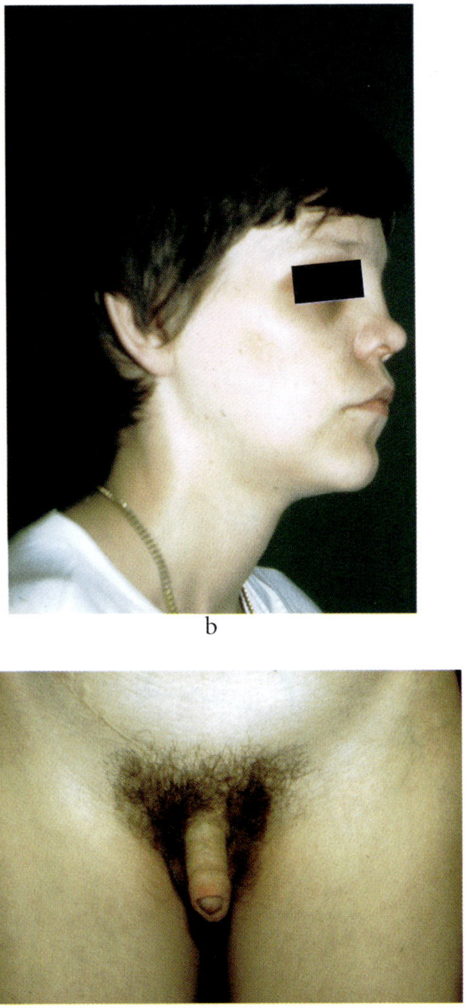

b

c

Figure 2.2a-c: A 25-year-old man with congenital secondary hypogonadism (Kallmann's syndrome), eunuchoid body proportions (over-long arms and legs with displacement of ratio of upper to lower body), absence of body and facial hair, horizontal delimitation of pubic hair, weak musculature, reduced testicle and scrotum size and androgen-deficiency anemia.

Once the body has taken on eunuchoid proportions the process is irreversible.The voice does not break as testosterone-induced growth of the larynx and elongation of the vocal cords does not occur. Growth of facial and body hair is minimal or completely absent. Pubic hair is sparse and delimited horizontally. In patients with hypogonadism prior to puberty, the volume of the testes usually remains infantile.

Adulthood

If the onset of hypogonadism does not occur until after puberty, the clinical picture can vary enormously. Body proportions, size of penis and pitch of voice do not change, but body and facial hair dwindle and shaving does not need to be carried out as frequently. The chief clinical signs are decreased sexual potency, loss of libido and infertility. Nevertheless, reduced libido and impotence do not cause any psychological suffering, unlike erectile dysfunction which is not endocrine-related. Size and consistency of testes varies enormously - ranging from normal to significantly reduced. Long-term androgen deficiency routinely causes severe osteoporosis, with lumbago and fractured bones which can be so bad that the patient is severely handicapped. Testosterone stimulates renal and extrarenal production of erythropoietin and division of erythropoietic stem cells. Therefore, testosterone deficiency often leads to normochromic, normocytic anemia with pallor, fatigue and neurasthenic fatigue. Loss of strength and vitality are also the result of atrophied muscle mass as the anabolic effect of androgen is absent (☞ Figure 2.3). As spermatogenesis requires the influence of testosterone, testosterone deficiency frequently leads to a reduction in sperm production, depending on the extent of the deficiency. Affected patients are then subfertile or infertile. Thus, the principal symptoms of androgen deficiency acquired in adulthood are decreased sexual potency, loss of libido and infertility. More rarely, men present with osteoporosis or extreme fatigue as the main symptoms.

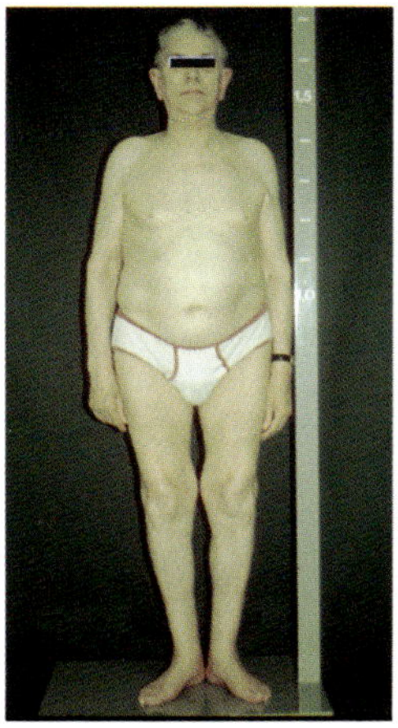

a

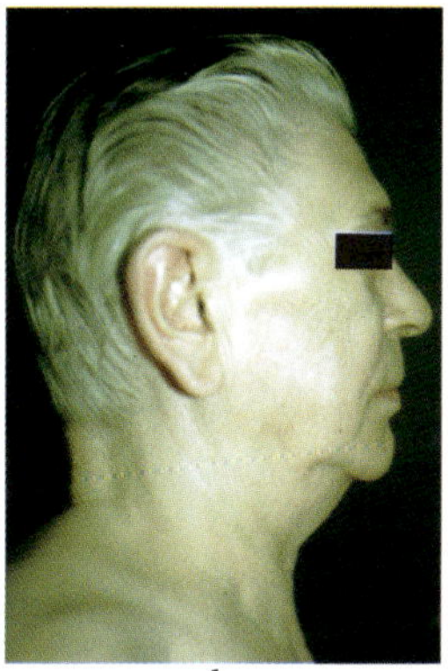

b

Figure 2.3a+b: A 60-year-old man with secondary hypogonadism acquired in adulthood as a result of a prolactin-secreting pituitary adenoma. Note the muscular atrophy, lack of body and facial hair and fine wrinkles of the facial skin, especially around the eyes. The slight gynecomastia is the result of hyperprolactinemia combined with testosterone deficiency. The patient himself reported loss of libido and sexual potency, as well as weakness and fatigue.

2.2. Causes and differential diagnosis of hypogonadism

The male reproductive system comprises four components: hypothalamus, pituitary, testis and androgen-sensitive target organs. Hence, disorders of any of the systems involved can lead to hypogonadism (☞ Table 2.2). Hypogonadism can be caused by a number of disorders. However, it is normally caused by very few. Even in a special endocrinology out-patient clinic, over 80 % of patients will be suffering from idiopathic hypogonadotropic hypogonadism, hypopituitarism or Klinefelter's syndrome.

Table 2.2 summarizes all the disorders that can cause hypogonadism. For the sake of completeness, disorders exclusively caused by an exocrine dysfunction of the testes (e.g. varicocele, germinal cell aplasia) are included in the list, but they are not mentioned in this book owing to its subject matter.

2.2.1. Hypothalamic male hypogonadism

The hypothalamus secretes the hormone GnRH in rhythmic pulses. GnRH stimulates the production and secretion of the gonadotropins LH and FSH in the hypophysis. Hence, hypothalamus-related causes of hypogonadism are characterized by reduced blood levels of LH, FSH and testosterone. This is called secondary hypogonadism.

2.2.1.1. Idiopathic hypogonadotropic hypogonadism and Kallmann's syndrome

Idiopathic hypogonadotropic hypogonadism (IHH) is caused by insufficient secretion of GnRH from the hypothalamus. As a result, LH and FSH are not produced by the hypophysis and therefore neither androgen production nor spermatogenesis are stimulated in the testes. In Kallmann's syndrome, there is additionally anosmia or hyposmia (partial or complete loss of sense of smell) for aromatic substances such as coffee or perfume.

Sporadic and familial forms of IHH have been described, where the different genetic modes (X-chromosomal, autosomal dominant with variable penetration and autosomal recessive) point to different pathogenetic mechanisms. IHH and Kallmann's syndrome occur in fewer than 1 in 8000 newborns, with a five-fold greater incidence in males. This makes the X-chromosomal inherited variant the most common form of IHH and Kallmann's syndrome[349].

Besides forms of GnRH deficiency where GnRH is completely absent, there are also variants where GnRH secretion is reduced. One rudimentary form is Pasqualini's syndrome which was formerly assumed to be due to a selective deficiency of hypophyseal LH. It now appears likely that Pasqualini's syndrome is due to reduced secretion of GnRH,

Site	Clinical picture	Cause	Androgen deficiency	Infertility
Hypothalamus (secondary hypogonadism)	idiopathic hypogonadotropic hypogonadism and Kallmann's syndrome	constitutional disorder of GnRH secretion	+	+
	Pasqualini's syndrome	LH deficiency due to disorder of GnRH secretion	+	(-)
	isolated FSH deficiency	FSH deficiency due to disorder of GnRH secretion	-	+
	Prader–Labhart–Willi syndrome	constitutional disorder of GnRH secretion	+	+
	Laurence–Moon–Bardet–Biedl syndrome	constitutional disorder of GnRH secretion	+	+
	familial cerebellar ataxia	constitutional disorder of GnRH secretion	+	+
	constitutional delayed puberty	'slow biological clock'	+	+
	secondary disorders of GnRH secretion	tumors, infiltrations, trauma, radiotherapy, malnutrition, vascular origin, drug abuse, medication, infection	+	+
Pituitary (secondary hypogonadism)	hypopituitarism	Infiltrations, adenoma, ischemia, empty sella syndrome, radiotherapy, postoperative, drug abuse, medication	+	+
	hyperprolactinemia	adenoma, medication	+	+
	biologically inactive LH	mutation in the LH gene	+	+
Testes (primary hypogonadism)	congenital anorchidism	loss of testes in the fetus	+	+
	acquired anorchidism	trauma, torsion, surgery	+	+
	pure gonadal dysgenesis	defect of the Y-chromosome (?)	+	+
	mixed gonadal dysgenesis	delayed development of the testes, synthesis disorder in testis of the fetus (?)	+	+
	oviduct persistence	lack of anti-Müllerian hormone	-	(-)
	germinal cell aplasia (Sertoli-cell-only syndrome)	constitutional or acquired (radiotherapy, medication, infections)	-	+
	Leydig cell aplasia	mutation in the LH receptor gene	+	(+)
	male pseudohermaphroditism	defective enzyme of testosterone biosynthesis	+	+
	Klinefelter's syndrome	aberration in number of chromosomes	+	+
	XYY syndrome	aberration in number of chromosomes	(+)	(+)
	XX-male syndrome	incomplete translocation of a part of the Y-chromosome	+	+
	male Turner's syndrome (Noonan syndrome, 45,X0)	incomplete translocation of a part of the Y-chromosome	+	+
	tumors of the testis	unknown	+	+
	varicocele	disturbed blood supply to the testis as a result of venous insufficiency	(-)	+
	orchitis	infection (viral, bacterial)	(-)	+
	globozoospermia	disorder of spermiogenesis	-	+
	biologically inactive FSH	mutation in the FSH gene	-	+

	non-motile cilia syndrome	disorder of spermiogenesis	-	+
	idiopathic infertility	unknown	-	+
	systemic diseases	e.g. renal insufficiency, hemochromatosis, liver cirrhosis, HIV infection, diabetes mellitus and many others	+	+
	exogenous toxins	medication, radiation, environmental toxins and substance and alcohol abuse	+	+
Afferent seminal duct, other sex glands	infections	bacteria, chlamydia, viruses	-	+
	obstruction	congenital malformation, infection, vasectomy	-	+
	Young's syndrome	unknown	-	+
	cystic fibrosis	mutation in CF gene	-	+
	liquefaction disorder	unknown	-	+
	immunologically induced infertility	autoimmune disease	-	+
Sperm deposition	deformed penis	congenital or acquired	-	+
	hypospadia or epispadia	congenital testosterone deficiency of the fetus	(+)	(+)
	phimosis	congenital	-	(+)
	erectile dysfunction	circulatory disorders, testosterone deficiency, neurogenic, psychological	(+)	(+)
Target organs of androgen	testicular feminization	defective androgen receptor gene with total loss of function	+	+
	partial testicular feminization	defective androgen receptor gene with widespread loss of function	+	+
	Reifenstein's syndrome	defective androgen receptor gene with moderate loss of function	+	+
	infertility with androgen resistance	defective androgen receptor gene with slight loss of function	-	+
	perineoscrotal hypospadia with pseudo vagina	mutation in 5α-reductase gene	+	+
	aromatase deficiency	defective aromatase gene	-	(+)
	estrogen resistance	defective estrogen receptor gene	-	(+)

Table 2.2: Summary of causes of hypogonadism in man. FSH, follicle stimulating hormone; GnRH, gonadotropin releasing hormone; LH, luteinizing hormone; HIV, human immunodeficiency virus; CF, cystic fibrosis.

enough to stimulate FSH but levels of LH are not completely normal. Nevertheless, the amount of LH secreted is sufficient to increase intratesticular concentrations of testosterone, and therefore, in conjunction with FSH, spermatogenesis is stimulated. On the other hand, extratesticular levels of testosterone are significantly reduced, resulting in hypogonadism, which is why these patients are referred to as 'fertile eunuchs'. Physical examination reveals insufficient virilization but testes are usually of normal size. Hormone tests on patients with Pasqualini's syndrome show reduced levels of LH and testosterone and normal levels of FSH. Long-term testosterone treatment can be given for the androgen deficiency. If a patient wants to father children, hCG can be given. Even in isolated cases where patients have no FSH but normal levels of LH and testosterone, there appears to be a functional disorder of GnRH secretion (defective pulse rate?). Human menopausal gonadotropin (hMG) is administered to treat their infertility.

Etiology (☞ Table 2.3)

Gene mutations on the X-chromosome (KAL gene) have been identified as the cause of X-chromosomal variants of IHH (Xp22.3)[147]. This gene encodes a cell adhesion protein, called anosmin, which is involved in the co-ordinated migration of GnRH-producing neurons and olfactory neurons. The GnRH-producing neurons are initially located in the region of the cribriform plate in the roof of the nose, and migrate during early fetal development, with the aid of the cell adhesion protein anosmin, to their final location in the hypothalamus. If anosmin is impaired by inactivating mutations, the GnRH-producing neurons do not come into contact with the hypothalamic-pituitary portal system and the hypophysis is not stimulated. As the Y-chromosome only carries one pseudo gene for KAL, men display only one allele of the KAL gene and are therefore more frequently affected by Kallmann's syndrome than women.

A further gene on the X-chromosome also appears to be involved in the function of the GnRH-producing neurons. Mutations and deletions of DAX-1 (dosage-sensitive sex reversal, adrenal hypoplasia congenital critical region on the X-chomosome gene 1, Xp21), in addition to congenital adrenal aplasia, may cause hypogonadotropic hypogonadism.

DAX-1 is selectively expressed in the hypothalamus, hypophysis, adrenal gland, gonads and gonadal structure during development. It appears to be a transcription factor with gene-suppressant action. DAX-1 encodes a protein which is 470 amino acids in size, is an orphan receptor (receptor without previously identified ligands) and is related to the family of steroid receptors. It is found together with steroidogenic factor 1 (SF-1), which is also involved in the regulation and expression of hormones which are active in sexual differentiation. Mutations in the gene of DAX-1, besides causing congenital adrenal hypoplasia, also cause hypogonadotropic hypogonadism. In contrast, gene duplications of DAX-1 cause male pseudohermaphroditism, presumably by suppression of the SRY gene (sex-determining region Y). DAX-1 appears to be involved at all levels of the hypothalamic-pituita-

Gene	Gene Locus	Phenotype	Location of CNS defect	Heredity
KAL	Xp22.3	HH, anosmia	hypothalamus	X-chromosomal recessive
DAX-1	Xp21	HH, CAH	hypothalamus, hypophysis	X-chromosomal recessive
Leptin	7q31.3	obesity, HH	hypothalamus	autosomal recessive
Leptin-R	1p31	obesity, HH	hypothalamus	autosomal recessive
GnRH	8p11	HH	hypothalamus	autosomal recessive
GnRH-R	4q21.2	HH	hypophysis	autosomal recessive
HESX-1	3p21.1	septo-optic dysplasia	hypophysis	autosomal recessive
LH-b	19q13.32	isolated LH deficiency	hypophysis	autosomal recessive
FSH-b	11p13	isolated LH deficiency	hypophysis	autosomal recessive
PROP-1	5q	hyposomia, HH, hypothyroidism	hypophysis	autosomal recessive

Table 2.3: Gene mutations that can cause hypogonadotropic hypogonadism (HH). CAH = Congenital adrenal hypoplasia.

ry-testicular axis, as mutations of DAX-1 can cause hypogonadotropic hypogonadism as well as disorders of spermatogenesis. Over 50 different mutations of DAX-1 have been described in men.

Moreover, phylogenetic analysis shows that some disorders of GnRH secretion must be traced back to autosomal genes. Isolated mutations were identified in the GnRH (8p11-p11.2) and GnRH receptor (4q21.2) genes. Within one family, the same mutation causes hypogonadism of varying levels of severity[35,230]. In some patients the mutation causes the GnRH effect to be weaker, although remaining activity is maintained. Affected patients show symptoms of hypogonadism which are more or less pronounced, and may only become manifest in delayed puberty leading to reduced masculinization.

In addition, leptin, the adipocyte hormone, influences the secretion of GnRH from the hypothalamus. This is probably indirectly mediated via the neuropeptide Y transmitter (NPY). Leptin is produced in the adipose tissue by adipocytes and signals fat mass to the hypothalamus via specific leptin receptors, essentially to influence eating behavior and intake of nourishment. In healthy people there is a very close correlation between the amount of adipose tissue and levels of leptin in the blood. At the same time, leptin signals the amount of adipose tissue to the GnRH neurons and indicates whether the nutritional status is adequate for reproduction. Very reduced levels of leptin (as in anorexia nervosa) cause reduced production of GnRH, and therefore gonadotropin production is not stimulated. Inactivating mutations of leptin or its receptor also cause extreme obesity and secondary hypogonadism. By administering leptin to leptin-deficient obese infertile rats, not only was their weight reduced, but fertility was also restored. It is also very likely that leptin is involved in initiating puberty. Despite intensive attempts, few mutations of leptin or its receptor have been identified in man. The main symptom is extreme obesity which begins in the newborn.

HESX-1, a member of the homeobox gene, is a transcription factor that is expressed in the brain and hypophysis of the embryo. Mutations in the HESX-1 gene cause septo-optic dysplasia, a syndrome which is characterized by panhypopituitarism, atrophy of the optic nerve and central nervous system (CNS) midline defects (e.g. agenesis of the corpus callosum).

Mutations of the β-chain genes of LH and FSH are presented in Chapter 2.2.2.3.

Diagnosis

In rare cases, hypogonadism is noticed in the newborn due to a micropenis or undescended testicle, which should be corrected within the first 2 years of life.

> Most patients seek medical advice when puberty fails to set in.

It is difficult to differentiate IHH from constitutional delayed puberty if there is no disorder of the olfactory senses to indicate Kallmann's syndrome. Hence, one should always ask the patient about his ability to smell, for example, coffee, soap and perfume. Unlike in delayed puberty, nightly pulses of LH are absent in the blood serum in IHH and Kallmann's syndrome. The GnRH test will often help a differential diagnosis: in constitutional delayed puberty 100 μg GnRH will usually trigger a significant rise in gonadotropins, whereas in IHH this is only possible after stimulating the pituitary with pulses of GnRH for several days (priming). Occasionally, priming must also be carried out in some patients with delayed puberty, to make a retrospective differential diagnosis.

Adults with untreated IHH display the complete picture of eunuchoidism (☞ Figure 2.2). Undescended testes or orchidopexy are often reported in the case history. The volume of the testes is significantly reduced, seldom exceeding 5 ml. Gynecomastia rarely presents. A differential diagnosis should always consider hypothalamic-hypophyseal masses, but eunuchoid body proportions indicate that hypogonadism has been present since puberty. Therefore, nuclear magnetic resonance imaging (MRI) of the hypothalamic and hypophyseal region should be carried out regularly. In Kallmann's syndrome there will often be aplasia or hypoplasia of the olfactory bulb and tract. A spermiogram will usually show azoospermia, resulting in infertility.

Discovery of the genetic defect of one of the proteins relevant to cell migration also explains the occasional link between Kallmann's syndrome and other constitutional disorders, e.g. synkinesis (involuntary movement mirroring a voluntary movement of the opposite extremity), cerebellar ataxia, central deafness, disorders of the eye muscles, renal aplasia, talipes cavus and cleft lip with cleft jaw and palate. In a more extensive study of 170 patients with IHH or Kallmann's syndrome, 100 % of patients with mutations in the KAL gene displayed anosmia, 85 % synkinesis and 31 % unilateral renal agenesis. In sporadic cases of IHH, synkinesis and renal aplasia were a predictive marker of a mutation in the KAL gene[311]. These signs should always be sought, as they make it easier to reach a differential diagnosis of constitutional delayed puberty.

Therapy

To compensate for the androgen deficiency, testosterone substitution is given. Undescended testes should have corrective surgery to increase the chance of fertility later on and prevent possible degeneration. The anosmia cannot be corrected. If the condition is treated before eunuchoid stature develops, the excessive growth in height can be prevented and normal male proportions attained. If the patient wishes to father children, spermatogenesis can be stimulated either with pulsatile administration of GnRH or a combination of hCG/hMG[283].

2.2.1.2. Delayed puberty

Definition

Delayed puberty is defined as the onset of puberty (gonadarche) at a later age than the average appropriate for the ethnic population of the same gender. In central European boys, puberty starts with growth of the testes at 11.2 ± 1.5 years of age[227]. Hence, if the testes are smaller than 4 ml by the age of 14, or no pubic hair has developed by the age of 15, the patient has delayed puberty[285]. This occurs in approximately 5 % of all boys.

Etiology

The differential diagnosis of delayed puberty is primary or secondary hypogonadism, or in very rare cases androgen resistance (☞ Table 2.2).

> However, the most common cause of delayed puberty, and actually a normal variant, is constitutional delay of development (CDD), caused by individual fluctuations in maturity of the hypothalamus.

For many boys, there will be a history of delayed puberty in the parents, usually the father.

Diagnosis

It is important to distinguish between CDD and primary or secondary hypogonadism in terms of prognosis and therapy. Primary hypogonadism and androgen resistance can be easily distinguished by measuring gonadotropin and testosterone levels (☞ Table 2.4). In primary hypogonadism the gonadotropins LH and FSH are raised, and testosterone reduced; in androgen resistance all three parameters are raised (☞ Figure 2.4).

It is often difficult to distinguish between hypogonadotropic hypogonadism and CDD, as both involve reduced basal levels of gonadotropins. Case history details such as an existing or previously undescended testicle, as well as other congenital abnormalities (e.g. cleft lip and cleft jaw and palate), plus evidence of deficiencies of other pituitary hormones indicate organically induced secondary hypogonadism. Hyposmia or anosmia indicates Kallmann's syndrome (see above). A typical family history of delayed onset of puberty makes CDD more likely. In many boys with CDD, bone age is retarded by more than a year. In rare cases, bone age may fall behind chronological age by up to 4 years. However, in CGD, status of puberty, height and bone age (= biological age) frequently match, and therefore there is a discrepancy between physical development as a whole and chronological age (☞ Figure 2.5).

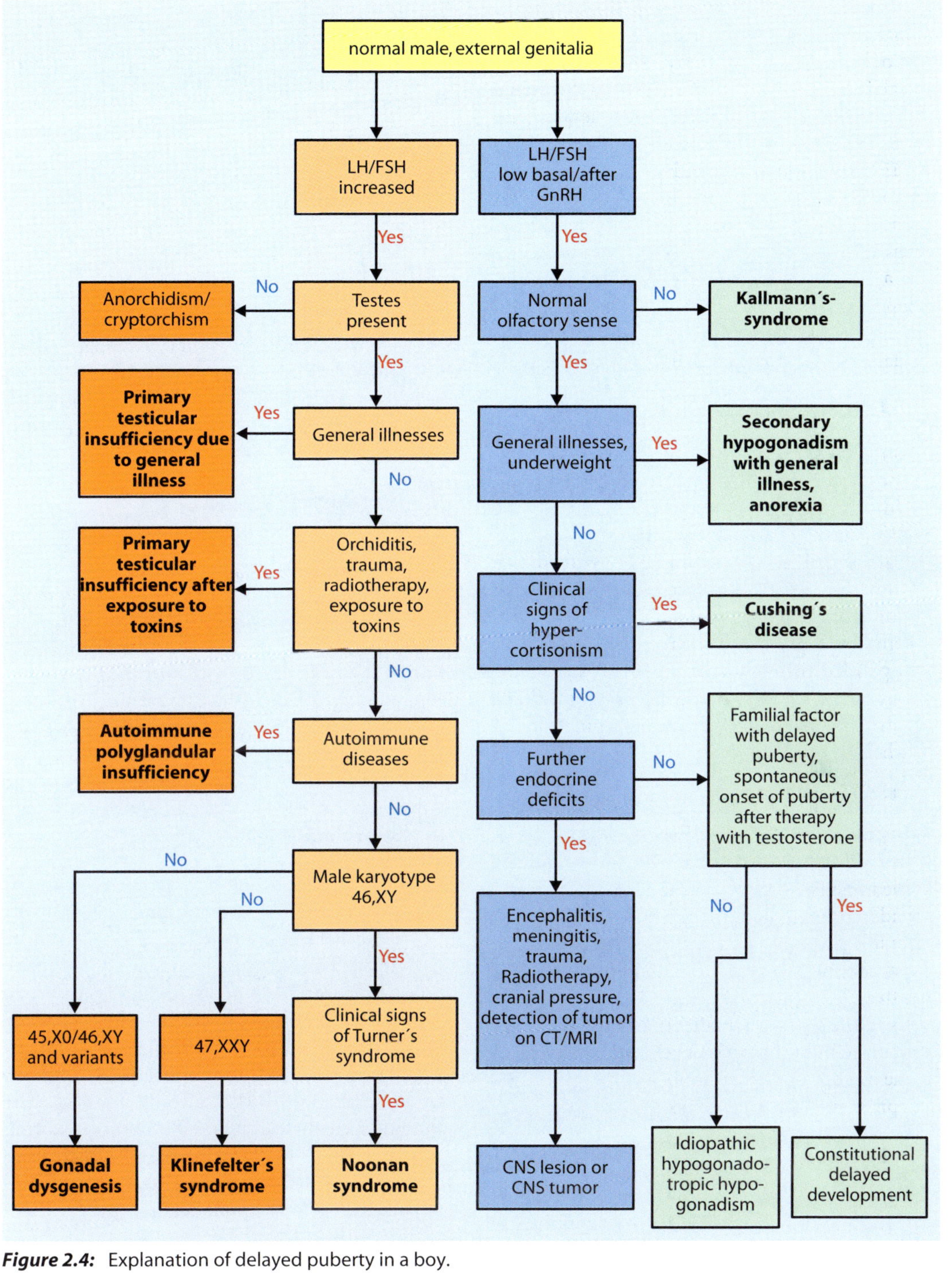

Figure 2.4: Explanation of delayed puberty in a boy.

Age	Testosterone (nmol/l)	Dihydrotestosterone (nmol/l)	LH (IU/l)	FSH (IU/l)
1-3 days	< 12	1.5-4.5	hCG	0-10
4-7 days	0.5-3.0	0.1-0.7	0-1	0-5
0.5-4 months	4-14	0.1-4.5	0-1	0-20
0.5-9 years	< 0.5	< 1	< 1	< 5
P1	0.1-1.0	0-0.2	0.5-2.5	0.5-3
P2	0.1-2.0	0.1-0.2	1-3	0.5-4
P3	0.3-12	0.2-0.9	1-4	2.5-4.5
P4	5-25	0.4-2.2	2-7	3-5.5
P5	10-32	0.6-3.5	2-7	2-5.5

Table 2.4: Normal ranges for testosterone, 5α-dihydrotestosterone and gonadotropins in young boys and adolescents. The reference ranges are dependent on the methodology used and therefore this table is only approximate. For P1-P5 see Table 1.3 and Figure 1.13.
LH = luteinizing hormone; FSH = follicle stimulating hormone; hCG = human chorionic gonadotropin.

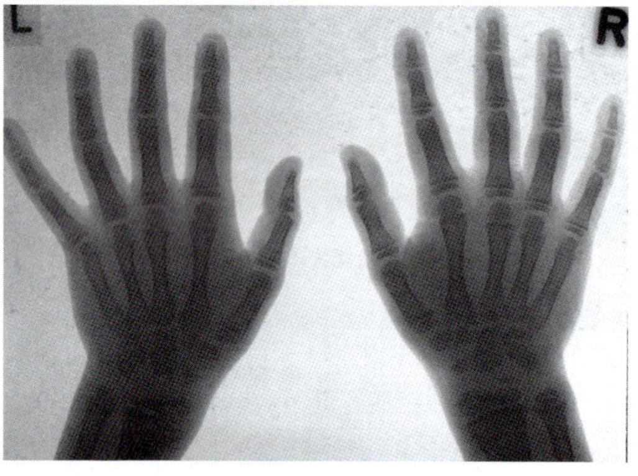

Figure 2.5: Open epiphyseal cartilage in a 16½-year-old boy with hypogonadotropic hypogonadism. Bone age is approximately 13½ years.

A significant increase in LH in the GnRH test (ΔLH > 10 IU/l) signifies CDD. However, the GnRH test may be negative in CDD as it is in hypogonadotropic hypogonadism. However, if one repeats the GnRH test after 36 h of pulsatile therapy with GnRH (5 µg intravenously every 90 min), it is possible to distinguish between CDD and hypogonadotropic hypogonadism in almost 100 % of cases. In CDD there is a greater increase in LH and less intense increase in FSH than in hypogonadotropic hypogonadism. Therefore, calculating the ratio of maximum FSH increase (ΔFSH) to maximum LH increase (ΔLH) will enable a distinction to be made[363]. In the case of secondary hypogonadism, nuclear MRI will have to be carried out to make a diagnosis.

Therapy

The most common cause of delayed puberty, CDD, may not need treatment, as puberty will start spontaneously. The necessity to treat will depend on the psychological well being of the boy. If there are psychological problems owing to the patient's child-like appearance, short-term testosterone therapy can be given. This should not be started before the patient is 14, as increasing levels of testosterone close the epiphyseal cartilage of the long bones, resulting in reduced final height[54,56].

Direct administration of testosterone is preferred over pulsatile GnRH therapy and injections of hCG because it is more practical, its effect is more rapid and it has been used extensively for many years. Three injections of 50-100 mg testosterone enanthate at intervals of 4 weeks rapidly cause the onset of masculinization and improve psychological well being. The dose administered depends on age, bone age and height already attained. The spontaneous course of puberty is observed during a treatment-free interval of 3-6 months. Increasing volume of testes and increasing levels of testosterone indicate the start of endogenous puberty.

In delayed puberty owing to secondary hypogonadism, the underlying disease is treated where possible. If an increase in gonadotropins is not expected (e.g. idiopathic hypogonadotropic hypogonadism, empty sella or permanent injury to the hypothalamus or pituitary), testosterone therapy is administered for virilization. Treatment should be managed by an experienced endocrinology clinic.

In primary hypogonadism, the testosterone increase will be insufficient, leading to incomplete virilization. Therefore, in this case too, puberty should be encouraged by generous amounts of testosterone at the appropriate time[280]. Androgen deficiency is indicated by raised levels of LH.

2.2.1.3. Underweight, anorexia nervosa

Underweight resulting from chronic disease (malabsorption syndrome in gastrointestinal diseases, chronic renal insufficiency, tumor cachexia, severe systemic diseases, etc.), excessive sport or a psychological disorder (anorexia nervosa) can lead to hypogonadotropic hypogonadism.

Etiology

Pathophysiologically, the secretion of GnRH is reduced, resulting in a deficiency of LH and FSH, which in turn causes testosterone production and spermatogenesis to stop. This may be due to a deficiency in the adipose tissue hormone leptin, which is produced by adipocytes when the amount of fatty tissue in the serum increases, and provides feedback to specific receptors in the hypothalamus of the degree of fat in the body. Leptin influences neurotransmitters which in turn exert an effect on GnRH secretion. If the person is considerably underweight, reduced leptin production provides feedback to the hypothalamus that body fat is inadequate for reproduction.

The diagnosis is based on the clinical picture of cachexia, or at least underweight and hypogonadotropic hypogonadism. The weight increase that follows correction of the underlying disorder normalizes GnRH secretion again, and therefore long-term therapy is usually not required in anorexia nervosa or other disorders in which weight loss is reversible (☞ Figure 2.6). Nevertheless, temporary testosterone therapy may be used to support the weight increase.

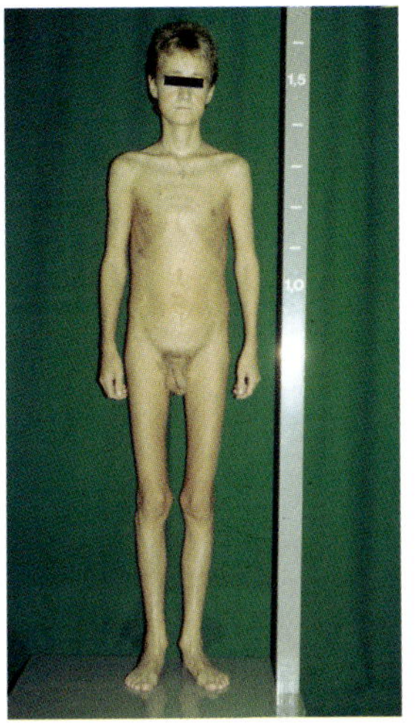

Figure 2.6: 23-year-old man with malabsorption resulting from Crohn's disease, severe underweight and hypogonadism due to this. This was corrected by weight increase alone in the context of treatment for Crohn's disease.

2.2.1.4. Prader-Labhart-Willi syndrome and Angelman's syndrome

Definition and etiology

This rare syndrome is characterized by a combination of the following symptoms: marked hypotonia (loss of muscular tonus), secondary hypogonadism, hyposomia, oligophrenia (mental deficiency), facial dysmorphism, obesity and type 2 diabetes mellitus. Prader-Labhart-Willi syndrome is caused by a deletion on the long arm of chromosome 15 (15q11-13). Recent molecular techniques have identified that the paternal chromosome 15 is always deleted in Prader-Labhart-Willi syndrome, whereas in Angelman's syndrome (oligophrenia, ataxic gait, dysmorphism), the mother's chromosome 15 is deleted. The incidence is estimated as 1: 20 000. Boys are over-represented.

Clinical picture and diagnosis

Hypogonadism is usually apparent at birth when the testes fail to descend and the penis is abnormally small (micropenis). It is caused by a disturbance in GnRH secretion from the hypothalamus, resulting in no onset of puberty. Obesity develops during childhood, and approximately 5-10 % of those affected by the marked obesity go on to develop diabetes mellitus. Diagnosis is based on the typical clinical picture, which frequently includes: acromicria, squint, scoliosis, hypoplasia of tooth enamel and a simian crease. Molecular genetic tests confirm the diagnosis, and prenatal diagnosis is also possible.

Therapy

There is no causal treatment available. In view of the reduced cognitive abilities and often severely abnormal behavior of the patient, testosterone or other drugs which increase testosterone synthesis (GnRH, hCG) should be administered with caution. Life expectancy is reduced by extreme obesity, which increases the risk of cardiovascular disease.

2.2.1.5. Laurence-Moon-Bardet-Biedl syndrome

The rare, autosomal-recessive inherited condition Laurence-Moon-Bardet-Biedl syndrome is characterized by five main symptoms:

- Hypogonadism
- Retinitis pigmentosa (tapetoretinal degeneration)
- Polydactylism/syndactylism
- Obesity
- Oligophrenia

In addition, patients frequently suffer from heart defects, parenchymal nephropathies and defects of the afferent urinary tracts, leading in 30 % of patients to terminal renal failure. As in Prader-Labhart-Willi syndrome, a dysfunction of the hypothalamus is thought to be responsible for the hypogonadism, although in some cases a disorder of the testicle has also been observed. Undescended testes, an abnormally small penis and hypospadia are frequently observed in the newborn. The hypogonadism is treated as in Prader-Labhart-Willi syndrome.

2.2.1.6. Rare syndromes

There are a number of very rare syndromes which cause secondary hypogonadism. These include familial cerebellar ataxia, which is associated with impaired hearing in the inner ear, cerebellar ataxia, short metacarpal bones of the fourth finger and hypogonadotropic hypogonadism[194].

2.2.1.7. Secondary GnRH secretion disorder as a result of other diseases

Any impairment of the hypothalamus can cause disorders of endocrine function (☞ Table 2.2). The clinical symptoms are determined by the location of the damage in the hypothalamus (affected brain centers), regardless of cause. If the preoptic region of the hypothalamus is affected, GnRH secretion may be disturbed, resulting in secondary hypogonadism, either as an isolated defect or in combination with other functional disorders of the hypothalamus. The cause of the impairment may be tumors (including craniopharyngioma, meningioma, glioblastoma) or granulomatous conditions (sarcoidosis, histiocytosis, tuberculosis or neurosyphilis). Causes such as medication or previous radiotherapy of the CNS may also cause impairment of GnRH secretion.

Clinical picture

In boys, the main symptom is delayed puberty since gonadotropin deficiency is only exposed at the time puberty is expected. Adults consult the doctor with the characteristic signs of androgen deficiency or infertility.

Diagnosis

The combination of reduced serum levels of testosterone and basal gonadotropins LH and FSH confirms secondary hypogonadism. An adequate increase in gonadotropins in the GnRH test distinguishes whether the hypogonadism is of hypothalamic or hypophyseal origin. Nuclear MRI is used to detect hypothalamic lesions. Pronounced masses can damage the optic nerves, and therefore sight and visual fields should be tested. If the damage to the hypothalamus is caused by trauma or radiotherapy, this should be identified in the case history. A complete endocrine status should always be made to exclude further endocrine deficits.

Therapy

Causal treatment is directed towards the underlying disorder. Delayed puberty is treated as outlined above. If androgen deficiency is to be corrected in the adult, testosterone substitution should be given. To achieve fertility, either hCG or hMG should be administered, or pulsatile GnRH if the hypophysis is intact[283].

2.2.2. Hypophyseal causes of male hypogonadism

The hypophysis produces the gonadotropins LH and FSH from the hypothalamus under the influence of the hormone GnRH. Inadequate secretion of these hormones causes hypogonadism as the testes are not sufficiently stimulated to function properly. Hormone analysis will typically reveal reduced levels of FSH, LH and testosterone. The GnRH test will reveal whether the disorder is related to the hypothalamus, in which LH and FSH will increase while there is no increase in the gonadotropins. Whereas gonadotropins increase in hypothalamic secondary hypogonadism (following repeat GnRH stimulation if necessary), no increase in gonadotropins is seen if the disordcr is hypophyseal in origin[181].

2.2.2.1. Hypopituitarism

Definition

Hypopituitarism is defined as insufficiency of the hypophysis, i.e. a deficiency of one or several pituitary hormones owing to various diseases.

Etiology

Any mass in the region of the hypophysis can lead to inadequate secretion of gonadotropins, and hence cause hypogonadism (☞ Table 2.5).

Masses
• Hypophyseal adenomas
• Residual cell tumors
- craniopharyngiomas, epidermoid tumors, Rathke's pouch tumors, chordomas
• Gamete tumors
- germinomas, teratomas, dysgerminomas, ectopic pinealomas
• Other tumors
- meningiomas, astrocytomas, gliomas, oligodendrogliomas, ependymomas
• Metastases
- bronchial carcinomas, carcinomas of the stomach, carcinomas of the breast, hypernephromas
Trauma
• Contusion
• Fracture of the base of the skull
• Disconnection of the hypophyseal stalk
Vascular causes
• Ischemia
• Postpartum pituitary necrosis (Sheehan's syndrome)
Iatrogenic measures
• Neurosurgery
• Cranial radiotherapy
Congenital defects
• Gene defects of receptors of hypothalamic releasing hormones
• Hypophyseal hormone
• Genes of hypophyseal structure and differentiation (PIT-1, PROP-1)
Granulomatous or inflammatory processes
• Autoimmune hypophysitis
• Sarcoidosis
• Tuberculosis
• Giant-cell granulomas
• Histiocytosis-X
• Abscess
Aneurysm
• Pseudo brain tumor
• Empty sella
• Arachnoid cysts

Table 2.5: Etiology of hypopituitarism.

The most common cause of hypophyseal insufficiency is benign adenoma of the anterior lobe of the hypophysis. Prolactinomas are by far the most numerous, followed by endocrine-inactive adenomas, and more rarely growth hormone-producing adenomas (acromegaly), adenocorticotropin (ACTH)-secreting adenomas (Cushing's disease), gonadotropin-producing adenomas and thyrotropin (TSH)-secreting adenomas.

Neurosurgical statistics show that pituitary adenomas account for 15 % of intracranial tumors. Microadenomas are distinguished from macroadenomas according to size (microadenoma < 10 mm, macroadenomas > 10 mm) (☞ Figure 2.7). Less common causes of hypophyseal insufficiency include intrasellar meningiomas, infiltrations due to sarcoidosis, hemochromatosis, histiocytosis X or infectious diseases such as tuberculosis and syphilis. Occasionally trauma to the skull or brain causes rupture of the hypophyseal stalk or circulatory disorders of the hypophysis, since the base of the brain does not have to be fractured. There have been very rare cases of 'empty sella', a congenital or acquired aplasia of the hypophysis.

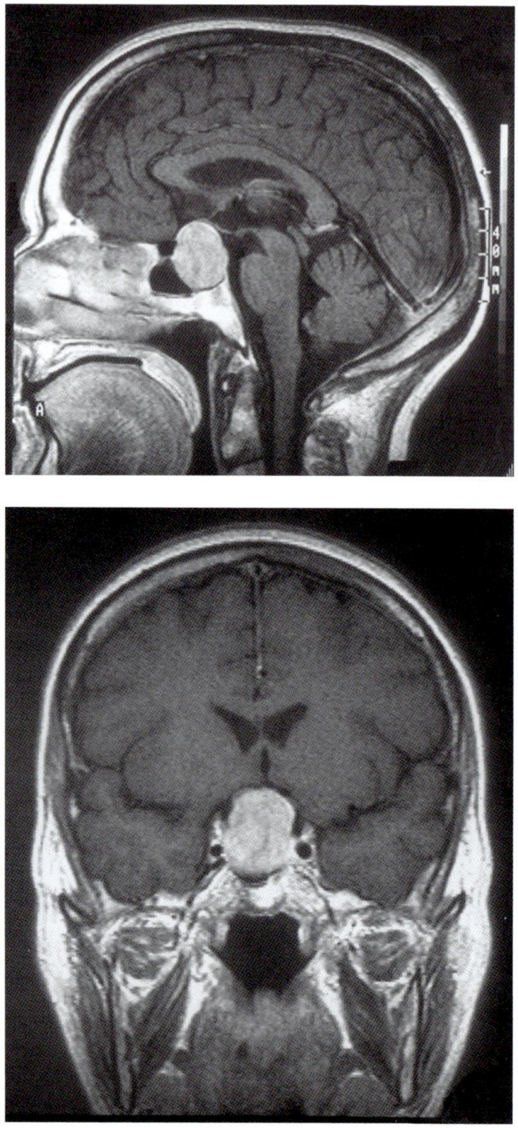

Figure 2.7: Macroadenoma of the pituitary, elevating the optic chiasm and parasellar extension adjacent to the carotids.

The transcription factors pituitary transcription factor-1 (PIT-1) and prophet of Pit-1 (PROP-1), which have only recently been identified, play a significant role in the ontogenesis, differentiation and function of somatotropic, lactotropic, thyrotropic and gonadotropic cells of the hypophysis. Mutations of the genes of these factors are increasingly described and lead to hypophyseal insufficiency with a variety of clinical symptoms. Patients with mutations in the gene for PROP-1 often display insufficiency of the anterior lobe of the hypophysis. Gonadotropins are not stimulated in the GnRH test. These patients display clinical signs which include delayed puberty. Other factors of hypophyseal differentiation include LHX3, a homeodomain transcription factor, and HESX-1, whose expression precedes that of PROP-1 (☞ Chapter 2.2.1.1). In addition, inactivating mutations of the LH β-chain have been described in some patients, leading to androgen deficiency and infertility (☞ Chapter 2.2.2.3).

Hypophyseal adenomas are classified according to function, size, invasivity and histology. The term endocrine-active hypophyseal tumors refers to hormone-producing hypophyseal tumors which cause clinical symptoms owing to an increased production of hormones. These include: prolactinoma, acromegaly, Cushing's disease, gonadotropinoma and thyrotropinoma. There are also mixed tumors which secrete more than one hormone and whose symptoms comprise a combination of the individual clinical pictures. This distinguishes these adenomas from clinically endocrine-inactive tumors either which do not actively secrete or whose product of secretion (e.g. α-subchain) does not cause symptoms despite increased serum levels. Classification according to histological dye behavior (eosinophil (growth hormone-producing), basophil (ACTH-producing), chromophobe (others)) has been superseded in this age of precise hormone diagnosis. More important are the dimensions, which are based on the largest diameter, and are divided into microadenomas (< 10 mm) or macroadenoma (≥ 10 mm). Classification according to intra-, supra- or parasellar extension, ideally according to Wilson's standard classification, is more informative[181].

The reason why hypophyseal adenomas develop is still unclear. Nevertheless, several studies have been able to show that almost all adenomas are monoclonal in origin, i.e. they originate from one cell. This indicates that changes (genetic?) in this original cell are responsible for the initial stages of development. Altered regulation of cell proliferation may be due to inactivating mutations of tumor-suppressant genes as well as activating mutations of oncogenes. Both processes have been observed in hypophyseal adenomas. However, apart from the gsp mutation in acromegaly and the menin gene in multiple endocrine neoplasia (MEN), no individual gene has been identified as being responsible for the development of hypophyseal adenomas.

Clinical picture

Disorders of the pituitary are often recognized from one or more of the three main clinical symptoms:

- Signs of increased hormone production
- Signs of reduced hormone production
- Signs of a mass

However, this simplified list hides a number of clinical symptoms. Some are presented in Table 2.6 by way of example. Whereas some symptoms are almost characteristic of a hormone disorder (e.g. growth of distal part of the extremities), other features represent non-specific symptoms with a number of differential diagnoses (e.g. neurasthenic fatigue or impaired erection). Moreover, the clinical symptoms of hormone disorders are also dependent on the age and sex of the patient as the possible causes show specific frequencies.

Absence of the gonadotropins LH and FSH is often the first indication of a hypophyseal disorder. In men, the main symptoms are those of androgen deficiency (☞ Table 2.1) and infertility.

Therefore, any hypophyseal diagnosis should be made with care. When using non-indicated hormone tests, pathological findings often result, which can be misleading and must be eliminated by extensive costly diagnosis. Not every newly discovered case of diabetes mellitus leads to a diagnosis of Cushing's disease. However, if there are further clinical symptoms present which suggest hyperadrenocorticism, appropriate hormone testing should be carried out, as no hypophyseal disorder should be overlooked. Table 2.7 lists important clinical findings relating to hypophyseal hormone disorders, to speed up diagnosis. Patients suspected of suffering from a hypophyseal disorder should always be referred to an endocrinologist for an optimal and rational diagnosis[181].

Diagnosis

Symptoms of androgen deficiency or infertility lead to testing for testosterone and gonadotropins, which are typically reduced. In the GnRH test, an inadequate rise in gonadotropins indicates the hy-

Main symptom	Symptom (selection)	Hormone responsible	Possible cause
Increased hormone production	galactorrhea	prolactin	prolactinoma
	growth of distal part of extremities	GH	acromegaly
	pronounced obesity of the trunk	ACTH	Cushing's disease
	weight loss	TSH	TSHoma
Reduced hormone production	growth impairment	GH deficiency	any mass in the sella region
	loss of libido/sexual potency, infertility, menstrual disorders	deficiency of gonadotropins	
	neurasthenic fatigue, hypotension	ACTH deficiency	
	weight gain, fatigue	TSH deficiency	
	neurasthenic fatigue	GH deficiency in the adult	
	agalactia	prolactin deficiency	
	polyuria/polydipsia	ADH deficiency	
Signs of a mass	headache	not hormone-related	any mass in the sella region depending on size
	visual impairment		
	reduced visual field		
	paralysis of muscles of the eye		

Table 2.6: Main symptoms of hypophyseal disorders based on typical examples.
GH = growth hormone; ACTH = adrenocorticotropic hormone; TSH = thyroid stimulating hormone (thyrotropin); ADH = antidiuretic hormone.

pophyseal origin of the secondary hypogonadism. A mass in the hypophysis will be detected by MRI. X-ray of the sella and lateral X-ray of the skull are now obsolete. An ophthalmological investigation should always be carried out to test the function of the optic nerves. A full endocrine diagnosis should accompany any impairment of hypophyseal function to ensure that all functional defects are discovered. Absence of the hypophyseal hormones ACTH and TSH in particular should not go unnoticed, as secondary hypothyroidism and secondary adrenocortical insufficiency can cause life-threatening complications.

Therapy

Therapy is based on the underlying disease leading to hypophyseal insufficiency. Whereas prolactinomas are almost exclusively treated with medication nowadays (see below), neurosurgical removal is indicated for other hypophyseal adenomas and sellar meningiomas. Ideally, this should be carried out in experienced neurosurgery clinics via the transnasal, transsphenoid route, to spare the healthy hypophyseal tissue.

Infections (meningoencephalitis, tuberculosis, syphilis) are treated with antibiotics or antivirals, and granulomatous infiltrations (e.g. sarcoidosis) with immunosuppressants. In autoimmune hypophysitis, treatment with corticosteroids can be attempted.

It is possible for hypophyseal function to return to normal if treatment of the underlying condition is successful. Therefore, secondary hypogonadism can be tolerated for a few months, especially if the hypophyseal microadenomas are easy to access. If recovery of hypophyseal function is unlikely or cannot be achieved in the short term, the hypogonadism should be compensated. Substitution of LH and FSH is necessary (to achieve fertility) only if a patient wants to father children[283]. Otherwise, only testosterone is substituted to compensate for the androgen deficiency. It is important to substitute other absent hypophyseal functions with cortisone (hydrocortisone or cortisone acetate), thyroxin and growth hormone, even in adults.

When the patient wants to have children, testosterone treatment is withdrawn and the testes stimulated with gonadotropins. In patients with hypothalamic hypogonadism and an intact hypophysis,

	Deficiency	Over-production
ACTH	**secondary adrenocortical sufficiency**	**Cushing's disease**
	• reduced exercise tolerance, neurasthenic fatigue, tendency to hypotension and hypoglycemia, wax-colored, often dry skin	• pronounced obesity of the trunk, buffalo hump, facies lunata, striae rubrae distensae, osteoporosis, pronounced proximal muscular weakness, tendency to hematoma, ecchymoses, hypertension, diabetes mellitus, acne, hirsutism, psychoses, menstruation disorders, loss of libido/sexual potency, infertility, increased skin pigmentation
LH/ FSH	**secondary hypogonadism**	**gonadotropinoma**
	• women: menstrual disorders, amenorrhea, infertility • men: loss of libido/sexual potency, infertility • children: delayed puberty	• usually asymptomatic • women: menstrual disorders, amenorrhea, infertility • men: loss of libido/sexual potency, infertility
GH	**GH deficiency**	**acromegaly**
	• reduced exercise tolerance, neurasthenic fatigue, tendency to hypoglycemia and hypercholesterolemia, reduced cardiac output and muscular energy, reduced bone mass, increased fat mass • children: reduced proportional growth, distal part of extremities petite, delayed growth of teeth, tendency to hypoglycemia	• growth of distal part of extremities, prognathism, large nose, supraorbital projection; headache, dental caries, protruding lips, macroglossia (enlarged tongue); deep, hoarse voice; struma, cardio- and visceromegaly, elongated colon; carpal and tarsal tunnel syndrome; secondary diabetes mellitus, hypertension, sleep apnea syndrome; arthrosis of the large joints, hyperhidrosis, thick often oily skin, hypertrichosis • children: gigantism
Prolactin	**prolactin deficiency**	**prolactinoma**
	• women: agalactia • men: no known sequlae	• women: menstrual disorders, amenorrhea, infertility, galactorrhea • men: loss of libido/sexual potency, infertility • children: delayed puberty
TSH	**secondary hypothyroidism**	**TSHoma (secondary hyperthyroidism)**
	• intolerance to cold, weight gain, fatigue, neurasthenic fatigue, slowing down, lethargy, depression, constipation, bradycardia, cool dry skin	• intolerance to heat, weight loss, agitation, mood swings, fine tremors of the fingers, tendency to diarrhea, tachycardia, moist warm skin

Table 2.7: Summary of most important clinical findings in patients with hypophyseal hormone production disorders.
ACTH = adrenocorticotropic hormone; LH = luteinizing hormone; FSH = follicle stimulating hormone; GH = growth hormone; TSH = thyroid stimulating hormone.

either pulsatile GnRH or human chorionic gonadotropin (hCG = LH activity) can be given in combination with human menopausal gonadotropin (hMG = FSH activity). Patients with hypophyseal insufficiency should only be given hCG/hMG (☞ Figure 4.11). Treatment is normally started with 1000-2000 IU hCG subcutaneously, twice weekly. As soon as serum testosterone levels are well within the normal range, an additional 150-225 IU hMG subcutaneously can be given three times a week. After 3-6 months an increase in size of the testes indicates stimulation of the germinal epithelium. Analysis of the ejaculate can start 6-9 months after treatment is initiated. An adequate sperm count is usually achieved after 12-24 months. However, with modern techniques of assisted reproduction with less than 100 000 sperm fertilization can be achieved.

2.2.2.2. Hyperprolactinemia and prolactinoma

Hyperprolactinemia describes any non-physiological increase in prolactin. Values above 20-25 ng/ml (400-500 mU/l) are deemed to be pathological. Prolactin reduces GnRH secretion from the hypothalamus, leading to secondary hypogonadism. An inhibitory effect on the hypothalamus is unlikely as there is usually a normal increase in gonadotropins in the GnRH stimulation test. Prolactin is not known to have a direct effect on the testes[181].

Etiology

There are many possible causes of hyperprolactinemia. Measuring the levels of prolactin in the serum will help to distinguish between them. If the values are more than 5-6 times higher than normal, prolactinoma is very likely to be the cause (☞ Figure 2.8). Then, a diagnosis can be made with the aid of nuclear MRI of the hypophyseal region, testing of the visual field and, if necessary, complete testing of endocrine pituitary function.

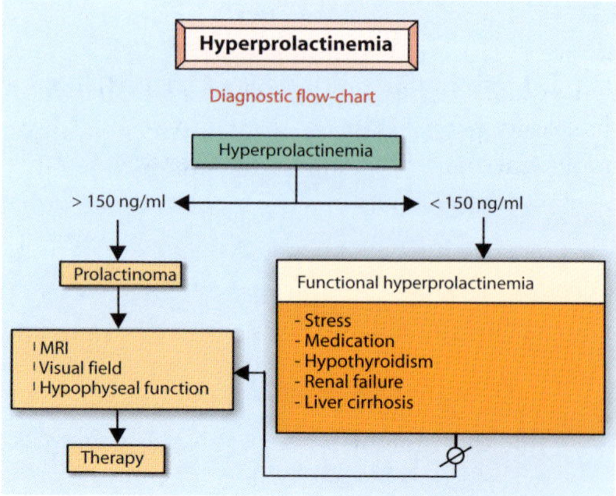

Figure 2.8: Flow chart explaining hyperprolactinemia.

Prolactinomas are normally monoclonal, i.e. they originate from one cell which, presumably due to a mutation, results in increased clonal proliferation and formation of a tumor. The monoclonal origin means that hypothalamic dysregulation (e.g. dopamine deficiency) is not responsible for inducing the prolactinoma. In some cases, changes in oncogenes in the dopamine receptor or the PIT-1 gene have been observed (myc, fos, jun, ras, gsp). However, a mutation has not been identified for the majority of prolactinomas. Unlike other hypophyseal adenomas, mutations of the regulatory regions of the G-protein have not been identified in prolactinoma. More recent findings indicate that nerve growth factor (NGF) and transforming growth factor α (TGF-α) are produced by lactotropic cells, and autocrine enhance the proliferation rate. The overexpression of TGF-α, which can be stimulated by estrogens, and the reduced effect of NGF, promote the development and growth of prolactinoma cells *in vitro* and in animal experiments. In addition, NGF stimulates the expression of dopamine-D2 receptors in lactotropic cells. This is essential for the response of dopamine to the hypothalamic inhibitory effect. NGF deficiency may therefore contribute to the development of an autonomic cell line[181].

Prolactinomas occur in approximately 20 % of patients with multiple endocrine neoplasia type I (MEN-I). The MEN-I gene (11q13) encodes a tumor suppressor, so that an inactivating mutation leads to the development of a tumor. As only a proportion of patients with mutations such as this also develop prolactinomas, at least one further modifying factor must be involved in the development of prolactinomas. Prolactinomas from MEN-I patients often show a more aggressive growth pattern than other prolactinomas. Prolactin-secreting hypophyseal carcinomas are rare. Fewer than 20 cases have been published to date.

Moderate increases in serum prolactin levels may be caused by hypothyroidism, renal insufficiency or liver disease. However, hyperprolactinemia is more frequently caused by medication, and therefore a detailed medication history should be taken (☞ Table 2.8). Slight increases in prolactin (less than twice the norm) are often an indication of stress, so before introducing further diagnostic measures prolactin levels should be repeatedly checked.

Antihypertensives	CNS drugs
α-Methyldopa, phenoxybenzamine, propanolol, reserpine, phentolamine	imipramine, Anafranil®, amitriptyline, chlorprothixene, chlorpromazine, pimozide, thioridazine, haloperidol, domperidone, meprobamate, chlordiazepoxide
Dopamine antagonists	**Hormones**
Metoclopramide	estrogens, thyrotropin releasing hormone
H_2-blockers	
Cimetidine, ranitidine	

Table 2.8: Drugs that may cause increased secretion of prolactin. CNS = central nervous system.

Clinical symptoms

Reduced libido and sexual potency as well as infertility are the main symptoms in man, and are almost always present. Approximately 30 % of men develop gynecomastia[65]. More rarely, galactorrhea occurs. If the patient has had secondary hypogonadism for some time owing to the hyperprolactinemia, the complete picture of hypogonadism with muscle atrophy, osteoporosis, androgen-deficiency anemia and sparse pubic hair develops (☞ Figure 2.3).

Diagnosis

Repeatedly high serum prolactin levels (men > 15-20 ng/ml, women > 20-25 ng/ml) are indicative of hyperprolactinemia. Further tests will enable a differential diagnosis to be made. Values of over 150-200 ng/ml are almost always associated with a prolactinoma. Values below 40 ng/ml are often stress-induced, and can be clarified by taking repeat blood samples. It is important to bear in mind that hyperprolactinemia is frequently physiologically or drug-induced (☞ Table 2.8). Suppression or stimulation tests are not suitable as they are not sufficiently accurate.

Therapy

In secondary hyperprolactinemia, the underlying condition is treated. In cases of hypothyroidism it is compensated, in drug-induced hyperprolactinemia, treatment should be adjusted. Stress-induced hyperprolactinemia does normally not require treatment.

> For patients with micro- and macroprolactinoma, dopamine agonists are now the drug of choice (☞ Table 2.9).

Owing to the orthostatic symptoms triggered by bromocriptine, 1 x 1.25-2.5 mg bromocriptine should be administered initially at bedtime for three nights and then gradually increased to the standard dose of 3 x 2.5-5 mg/day. It is possible to administer higher doses, but these are not always well tolerated. Newer dopamine antagonists, e.g. cabergoline and quinagolide, usually have fewer side-effects and are better tolerated than bromocriptine. The maintenance dose represents the minimum dose required to keep serum prolactin within the normal range. Initially high doses can often be reduced in time to very low doses without the prolactin increasing again. Some 90 % of patients on this therapy attain normal prolactin levels within a few months[89]. In over 80 % of patients the adenoma shrinks by more than 25 % of its original volume. Most of this shrinkage occurs in the first few months of treatment, but the adenoma will continue to shrink for years with treatment. There have even been reports of the adenoma disappearing completely. After 3-4 years of continual therapy with dopamine agonists and effective suppres-

Drug	Potency	Brand name	Administration (standard dose)
Metergoline	0.4	Liserdol®	3 x 4-8 mg/day
Dihydroergo-criptine	0.5		
Bromocriptine	1	Pravidel®	3 x 2.5-5 mg/day
		Kirim®	3 x 2.5-5 mg/day
		Parlodel LAR®	50-100 mg intramuscularly every 3-4 weeks
Terguride	7		
Cabergoline	10	Dostinex®	0.5-1 mg once or twice/week
Lisuride	12.5	Dopergin®	2-3 x 0.2 mg/day
Pergolide	100	Parkotil®	3 x 0.05-0.25 mg/day
Quinagolide	100	Norprolac®	1 x 75-150 μg/day

Table 2.9: Dopamine agonists for the treatment of prolactinoma. Typical side-effects of dopamine agonists which often occur during the initial treatment phase include: orthostatic hypotension, dizziness, nausea and dry mouth.

sion of prolactin, a withdrawal trial can be attempted. Approximately 10-20 % of patients will show normal levels of prolactin. A relapse will be indicated by rising serum prolactin levels after 2-3 months. Then therapy can be restarted at the dose last taken.

Macroprolactinomas present a significantly more demanding problem than microprolactinomas because of the hypophyseal insufficiency and chiasma syndrome which are often present, which is why they should be dealt with at a specialist endocrinology clinic. Patients with acute visual defects should be quickly started on high-dose dopamine agonist therapy. This should lead to an improvement in the visual disorder within days as the tumor reduces in size[251]. Treatment is closely monitored by ophthalmological tests and determination of prolactin levels. Macroprolactinomas without acute ophthalmological symptoms are treated in the same way as microprolactinomas (in an outpatient clinic with oral dopamine agonists). A meta-analysis of macroprolactinomas treated with dopamine agonists revealed shrinkage of the adenoma by more than 25 % in 215 out of 271 patients (79 %)[38]. The adenoma usually starts to shrink rapidly within the first few weeks or months, and then continues to shrink over subsequent years. A significant suppression in prolactin usually precedes shrinkage of the tumor. Macroprolactinomas are rarely completely eliminated by dopamine agonists, which is why therapy has to continue for a long time. Alternatively, radiotherapy or neurosurgery can be considered during dopamine agonist therapy. Surgery is more successful after prior treatment with dopamine agonists, presumably as a result of the tumor shrinking as this increases the chances of removing it completely[298]. If dopamine agonists are unlikely to control the growth of the adenoma sufficiently, the patient will be offered a neurosurgical operation. As the adenoma continues to be active postoperatively, radiotherapy is required by 60 % of patients to prevent the tumor growing back.

2.2.2.3. Selective deficiency of biologically active LH or FSH

Molecular techniques have enabled the identification of individual patients who produce structurally defective LH or FSH[302,426]. To date, only one patient has been found with a homozygotic mutation in the β-chain of LH (position 54, glutamine to arginine), which prevents binding to the specific LH receptor[426]. The boy had delayed puberty and low levels of testosterone. Although previous concentrations of LH measured by immunoassay were raised, the bioassay revealed an absence of biological activity. Administration of exogenous LH or hCG normalized testosterone production. Interestingly, the external genitalia of the boy were those of a completely normal male, although there must have been a deficiency in biologically active LH in the embryo. Either androgen production from the adrenal gland was sufficient *in utero*, or, more probably, androgen production from the testes was stimulated by the mother's hCG. In contrast, inactivating mutations of the LH receptor always cause aplasia or severe hypoplasia of the Leydig cells, resulting in male pseudohermaphroditism (☞ Chapter 2.2.3.8.2).

Inactivating mutations of the FSH β-chain were described in two men who had azoospermia[237,302]. In addition to the azoospermia, one patient had delayed puberty with low testosterone and completely normal LH, and small testicles resulting from lack of stimulation of the germinal epithelium. Molecular biological examination revealed a deletion (Val61X) in the coding region of the gene for the β-subunit of FSH, with consequent loss of the last 51 amino acids in the C-terminal end of the polypeptide chain. Since Leydig cells do not carry receptors for FSH, this surprising observation indicates an interaction between Sertoli cells and Leydig cells. The Sertoli cells appear to produce a so far unidentified substance under the influence of FSH, which stimulates the synthesis of testosterone in the Leydig cells and is imperative for the production of normal quantities of testosterone. Inactivating mutations in the gene of the FSH receptor cause subfertility or infertility in men, which cannot be treated[357,358]. Spermatogenesis may remain very variable depending on the the remaining activity of the mutated FSH receptor. Two out of five patients with FSH receptor mutations fathered children.

2.2.3. Testicular causes of male hypogonadism

The testes are bifunctional organs because they produce androgens and sperm.

Hypogonadism may therefore manifest in either reduced production of androgen or disturbed spermatogenesis, or a combination of the two. Disorders of the testes are described as primary hypogonadism, and differentiated according to whether they are hypothalamic or hypohyseal in origin. They are typically characterized by increased gonadotropins and reduced testosterone.

2.2.3.1. Congenital and acquired anorchidism

Definition, etiology

Anorchidism describes the absence of a testicle. Congenital bilateral anorchidism is rare, occurring in approximately 1 in 20 000 newborn boys. Unilateral anorchidism (monorchidism) is approximately four times as common as bilateral anorchidism. As boys affected with this condition display completely normal male sexual differentiation, endocrine active testicular tissue must have been present until at least the 14th week of pregnancy. Intrauterine torsion of the testes, vascular occlusion or intrauterine viral infections and teratogenic toxins are suspected to cause atrophy of the testis[36].

Acquired anorchidism is mainly due to removal of a testicle in case of a tumor. In rare cases it results from trauma, severe inflammations, torsion and operations that have gone wrong (herniotomy, orchidopexy).

Clinical symptoms and diagnosis

In a newborn phenotypical boy, the empty scrotum and the lack of evidence of testes in the inguinal canal on palpation and sonography lead to the suspicion of anorchidism. Occasionally an abnormally small penis will testify to the absence of testosterone stimulation in the later fetal stage. The male karyotype (46,XY) and the absence of oviducts (uterus, Fallopian tubes) enable differentiation from a completely virilized girl, e.g. with adrenogenital syndrome. The most important differential diagnosis of cryptorchidism (undescended testicle) can be made with the hCG test (5000 U hCG/m^2 of body surface area given intramuscularly, measurement of testosterone before and 72 h after administration). If testicular tissue is present, there will be a significant increase in serum testosterone to 1.5-2-fold the base value; this will not occur in bilateral anorchidism. Further development of boys with anorchidism is normal for their age until puberty. However, boys whose bilateral anorchidism has so far been undetected are then taken to the doctor with delayed puberty. The gonadotropins which are already raised before birth increase further at puberty, and these, combined with the low level of testosterone, raise suspicion of primary hypogonadism. In adolescents too, undescended testicles should be investigated with the hCG test, imaging techniques (sonography, MRI) and even laparoscopy, as these should be removed owing to the high risk of malignant degeneration.

Therapy

In bilateral anorchidism, testosterone substitution should be initiated when puberty is expected to start. This must be continued throughout a patient's life. To prevent premature closing of the epiphyses with reduced final height, testosterone substitution should be carefully initiated and supervised by an experienced endocrinology clinic. The infertility cannot be corrected. For psychological reasons, prosthetic testicles should be inserted in the scrotum. If the testes were removed to treat carcinoma of the prostate or sexual delinquiency, no androgen substitution is given.

Monorchidism does not require treatment as the remaining testicle completely fulfils the endocrine and exocrine function.

2.2.3.2. Malpositioning of the testes

Definition

If a testicle is located outside the scrotum, this is defined as abnormal. Table 2.10 lists the various forms and their frequency of occurrence.

Some 2-3 % of all mature, male newborns carried to term suffer from undescended testicles. In two-thirds the testicles descend spontaneously within the first 3 months. Thus, the prevalence of this condition is approximately 0.8 % in untreated 1-year-old boys. On the other hand, 30 % of premature boys suffer from undescended testicles. However, the testicles of most premature babies migrate spontaneously into the scrotum within a few months. All forms of unilateral undescended testicle occur five times more frequently than bilateral forms.

Etiology and pathophysiology

In healthy boys, the testicles migrate during the third trimester of pregnancy from an intra-abdominal position through the inguinal canal to their final position in the scrotum. The completely descended testicle is therefore one of the assessment criteria for the maturity of a newborn. Androgens promote the migration of the testicles[188]. This explains the increased incidence of malpositioning in all disorders of androgen production or androgen effect which become manifest in the fetus (e.g. forms of primary and secondary hypogonadism, male pseudohermaphroditism). In addition, intra-abdominal pressure appears to be significant to the descent of the testicle, as defects in the abdominal wall (exstrophy of the bladder, umbilical hernia, gastroschisis) or hypotension of the abdominal muscles (prune belly syndrome) also cause malpositioning (of the testicle). Furthermore, undescended testicles occur in many congenital syndromes[194]. However, in 85 % of patients no disorder is detected, and hence the malpositioning is purely idiopathic.

Malpositioning of the testes causes irreversible damage to the germinal epithelium, presumably because of the higher temperature. This can be detected in biopsies of the testes in the first year of life, and becomes progressively worse, the longer the testicles are in an abnormal position[141]. However, further factors in addition to a rise in temperature are significant, as in unilateral undescended testicle the contralateral, descended testicle often also shows disturbed spermatogenesis. Endocrine

Form	Malpositioning (% frequency)	Definition
Pendular testicles	72	the testicle moves between the scrotal and inguinal position; this condition is caused by cremasteric reflex caused by cold stimulus or coitus
Sliding testicles	15	the testis lies on the external inguinal ring and can only be pressed into the upper scrotal space upon application of pressure but then slips back again spontaneously
Retained testicle	10	the testis lies fixed in the inguinal canal
Cryptorchidism (undescended testicle)	< 3	the testis lies above the internal inguinal ring and can neither be seen or felt
Ectopic testicle	< 0.5	the testis lies outside the normal route of descent, e.g. femoral or perineal

Table 2.10: Definitions, distribution and characterization of abnormal testicular positioning.

function usually remains intact in idiopathic malpositioning of the testicle[188].

Clinical picture

In the newborn, undescended testes should be determined in the preliminary examination. To differentiate between pendular and sliding testicles (the former cannot be treated, the latter always can) they can be investigated in a warm bath. Pendular testicles will slide spontaneously into the scrotum when the cremasteric muscle relaxes. Inguinal testes are identified by palpation and sonography. If neither testicle is visible, an hCG test is performed to distinguish between cryptorchidism and anorchidism. If only one testicle is missing, this should be sought by ultrasound or, if necessary, MRI.

Men often consult a doctor for the infertility caused by the undescended testicle. Semen analysis usually reveals oligoasthenoteratozoospermia or, less commonly, azoospermia. Testosterone and LH are usually within the normal range, whereas FSH is often increased owing to the irreversible damage to the germinal epethelium.

Malpositioning increases the risk of developing testicular tumors by ten-fold. The incidence of testicular tumors is higher in cryptorchidism than in retained or sliding testes. If the abnormality is unilateral, the contralateral testicle also has an increased risk of developing a tumor. The most common tumor in undescended testes is seminoma, which is why undescended testicles should be regularly checked by sonography and palpation even after the condition has been corrected.

Therapy

It has been histologically proven that the germinal epithelium can become damaged within the first 12 months of a baby's life, so to avoid reproductive problems in the adult, attempts are now made to return the testicle to its position in the scrotum by the end of the first year. Early treatment may reduce the likelihood of developing testicular tumors later on.

If the testicle has not spontaneously descended by the third month, treatment is initially attempted with intramuscular hCG or intranasal GnRH. The success rate is the same for both forms of treatment - somewhere between 10 and 50 %[312]. A dose of 2 x 250 IU hCG is administered per week for 5 weeks until the child is 1 year old and then 2 x 500 IU weekly until the child is 6. Older children receive 2 x 1000 IU hCG. If necessary, a further cycle of treatment can be given after 3 months.

Intranasal administration of 200 µg GnRH (all age groups) into each nostril as a spray three times daily for 4 weeks can be more pleasant for children than intramuscular injections, and compliance is good. If the testicle remains undescended despite hormone therapy, orchidopexy should be performed without delay. If there is a hernia in addition to malpositioning, orchidopexy should be performed in conjunction with a herniotomy without prior drug treatment.

There is no drug that will treat fertility disorders caused by undescended testicles, as the germinal epithelium is irreversibly damaged. Depending on the findings from semen analysis, assisted reproduction techniques may be attempted (e.g. *in vitro* fertilization).

If an adult has a malpositioning condition, this should be corrected by the time he is 32 years old. Attempts at treating with hCG or GnRH after this age have been unsuccessful. If the undescended testicle is unilateral, particularly if it lies in the inguinal or abdominal position, unilateral orchiectomy is recommended; if malpositioning is bilateral, both testicles should be moved into the scrotum to aid palpation and sonography[107]. In men over 32 years of age, the testes are left where they are, as the risk of developing testicular tumors decreases significantly after this age; nevertheless, regular checks must be carried out. In rare cases, endocrine insufficiency with androgen deficiency will develop with idiopathic malpositioning, and substitution treatment must then be given.

2.2.3.3. Orchitis

Definition and etiology

Orchitis is very rare nowadays, and usually the result of a viral infection (mumps virus, ECHO (enteric cytopathogenic human orphan-viruses, arboviruses, varicella (chickenpox), Coxsackie A, Marburg virus). Orchitis due to mumps is the most common viral infection of the testes, and occurs in approximately 25 % of adolescents or adults with parotitis. In a third of these patients both testes are affected. Prepubescent boys practically never suffer from orchitis.

Epididymitis or even orchitis can arise following bacterial infections such as urethritis, cystitis or prostatitis if the pathogen ascends further up the tract. Orchitis due to gonorrhea or tuberculosis is very rare today. Infertile men are very commonly found to be suffering from mycoplasma infections, although no causal connection has been confirmed.

Clinical picture

The acute inflammation associated with viral orchitis is characterized by painful swelling of the testicles, nausea, headache and fever. Acute testicular insuffiency is revealed by reduced levels of testosterone and increased levels of LH and FSH in serum. As the inflammation goes down testicular volume also reduces. Orchitis due to mumps may be preceded by parotitis. This renders differential diagnosis difficult, although it may be possible to make a diagnosis by determining specific immunoglobulin M (IgM) antibodies. If there is no atrophy of the testes, testosterone and LH return to their original values after the inflammation has subsided.

Histologically, there is edema initially with mononuclear infiltration of the interstices and gametes. Later, there is degeneration of the gametes, which develops into total germinal aplasia and sclerosis of the tubules. There is a sharp increase in intratesticular pressure, resulting in ischemia. This, together with the virus itself, may be responsible for the irreversible degeneration of the germinal epithelium. Accordingly, bilateral orchitis due to mumps often causes pronounced oligospermia or azoospermia with subfertility or infertility, which cannot be treated with medication. FSH is often raised as a result of the damage to the germinal epithelium. Following unilateral orchitis, the contralateral testicle occasionally shows histological changes. A subclinical virus or immunological processes are suspected to be the cause. Androgen deficiency rarely results.

If a bacterial infection is suspected, in addition to microbiological investigation of a urethral swab, it is sensible to culture the organisms and draw up an antibiogram-resistogram of the organisms in the semen.

It is important to make a differential diagnosis of torsion of the testes, as surgery performed at an appropriate time may prevent irreversible damage.

Therapy

During the acute inflammatory stage of viral orchitis, the scrotum should be elevated and kept cool, and corticosteroids administered (e.g. 50 mg prednisolone for 2-3 days then gradually reducing the dose). This will reduce swelling and relieve pain. If necessary, analgesics should be administered. There is no treatment for the infertility caused by orchitis. If testosterone deficiency should arise, which is rare, testosterone substitution should be given.

Bacterial orchitis or epididymitis can be specifically targeted following an antibiogram. However, before these findings are known, a gyrase inhibitor can be administered and, if response is good, treatment will not need to be changed. In epididymitis in particular the vas deferens can become obstructed by adhesion, and this may cause infertility, which is why bacterial infections should be treated as quickly as possible.

2.2.3.4. Klinefelter's syndrome

Definition

Klinefelter's syndrome is a congenital aberration of the number of chromosomes and, in the classic case, has the karyotype formula 47,XXY.

Etiology

The condition is caused by one or more extra X-chromosomes as a result of non-separation during the first meiotic division of the parent gametes (☞ Figure 3.4). The extra X-chromosome more commonly comes from the mother than from the father and the risk increases with age. In fewer than 5 % of patients, the non-disjunction occurred during mitosis of the zygote. Some 80 % of patients display the classic karyotype 47,XXY. Other men with Klinefelter's syndrome have variants which may be also due to extra Y-chromosomes (e.g. 48,XXXY, 48,XXYY).

Unlike many other numeral aberrations of chromosomes, Klinefelter's syndrome is not associated with an increased rate of miscarriage. It is not known why the additional X-chromosome impairs testicular endocrine and exocrine function. Pseudoautosomal genes on the X chromosome (e.g. adenosine diphosphate/adenosine triphosphate (ADP/ATP)-translocase, Xp22.3) may play a role which is beyond the normal X inactivation in the presence of two X-chromosomes[334].

Clinical picture

The clinical picture is characterized by a combination of very small testicles, infertility, gynecomastia and hypergonadotropic hypogonadism and becomes apparent during puberty (☞ Table 2.11)[187]. The signs of Klinefelter's syndrome are almost unnoticeable in childhood. Occasionally boys affected with the condition are referred for hypoplasia of the external genitalia or extra-long legs (☞ Figure 2.9).

Karyotype	47,XXY	46,XY/47,XXY
Pathological testicular histology	100	94
Small testicles	100	73
Azoospermia	100	50
Androgen deficiency	79	33
Gonadotropins increased	75	33
Reduced facial hair	61	64
Reduced libido or sexual potency	30	56
Reduced or feminized pubic hair	41	62
Small prostate	46	18
Gynecomastia	76	33
Small penis	10	21

Table 2.11: Frequency (%) of characteristic symptoms of Klinefelter's syndrome in 692 men with classic karyotype (47,XXY) or mosaic form[28,127].

If puberty starts when it should, virilization is usually sufficient. However, the testes do not grow at all, and their volume seldom exceeds 4 ml. Histological examination indicates a hyaline degeneration of the germinal epithelium with very few intact gametes. Therefore, the ejaculate almost always shows azoospermia. Individual tubules with intact spermatogenesis are seen in patients with the mosaic form 46,XY/47,XXY, and therefore very occasionally motile sperm are found in the semen[187]. The total number of Leydig cells appears to be reduced.

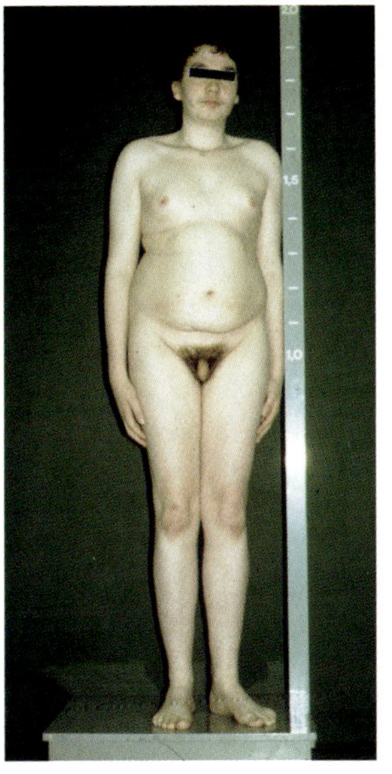

Figure 2.9: A 21-year-old man with Klinefelter's syndrome (47,XXY). Note the lack of secondary hair (horizontal pubic hair line, absence of hair on the trunk and legs, almost absent facial hair), gynecomastia and female-type fat distribution with pronounced hips. The lower body is significantly longer than the upper body.

Libido and sexual potency are initially normal, but decrease between the ages of 25 and 35. This reflects the increasing insufficiency of the Leydig cells. Without androgen substitution the typical clinical signs of androgen deficiency develop.

Patients with Klinefelter's syndrome are often of greater than average height. This is due to the increased growth of the long bones of the legs (length of lower body > length of upper body). Factors other than hypogonadism appear to play a role, as the increased arm span which is typical of eunuchoidism is absent in Klinefelter's syndrome.

Over one-half of patients suffer from gynecomastia, which is sometimes considerable. The development of enlarged breasts starts in puberty, and is presumably a consequence of the androgen-estrogen balance which has shifted towards estrogen[65]. In addition, the incidence of mammary carcinoma is significantly higher in patients with Klinefelter's syndrome. Approximately 3-5 % of all Klinefelter patients develop breast cancer[105,336]. The risk increases if there is a family history of mammary carcinoma among female relatives. Extragonadal gamete tumors also occur more frequently than usual. Young patients up to 25 years old are predominantly affected.

Varicosis of the lower leg and chronic venous insufficiency occur exceedingly frequently in patients with Klinefelter's syndrome[187].

Most, but not all, patients with Klinefelter's syndrome are of lower intelligence[279]. Whether genetic causes are responsible, or the premature recognition of being different leads to a feeling of isolation at school with consequent poor learning, is unclear. The majority of patients with Klinefelter's syndrome lead a completely normal social and married life.

Diagnosis and laboratory findings

A testis volume of less than 5 ml should always lead to a suspicion of Klinefelter's syndrome. It can be differentiated from secondary hypogonadism by increased serum concentrations of LH and FSH. Levels of FSH will usually be much higher (☞ Figure 2.10).

Serum testosterone is often reduced, or lies in the lower normal range (40 % of patients). Free testosterone is more frequently reduced than total testosterone as the transport protein sex hormone-binding globulin (SHBG) is increased. As LH stimulation is increased, the Leydig cells produce more estradiol in relation to testosterone secreted, and therefore serum estradiol concentrations are occasionally increased. The increased estrogen-induced production of SHBG shifts the androgen-estrogen balance even further towards estrogen.

The Barr chromatin body test (determination of extra X-chromosomes) is a rapid test used to determine chromosomal abnormality in a swab of cheek mucosa. Even if the findings are negative, the lymphocytes should be karyotyped, as not all cell lines in the mosaic forms show the extra X-chromosome. This is why further tissues such as skin fibroblasts or, less commonly, when there is an urgent request for an accurate diagnosis, testicular tissue, must be karyotyped.

Therapy

There is no causal treatment for the chromosome disorder and the infertility. However, it is possible today, using modern techniques of assisted reproduction, to help patients with Klinefelter's syndrome become fathers[146]. Injection of sperm, obtained either from isolated semen or through aspiration from the rete testis, into egg cells, and the subsequent transfer of embryos, have successfully led to numerous full-term pregnancies and the birth of healthy children. Nevertheless, there is still the risk that an extra sex-chromosome may be transferred to the zygote, and therefore it is sensible to give the couple genetic counselling and to perform a chromosome analysis of the blastocysts[118,437]. A

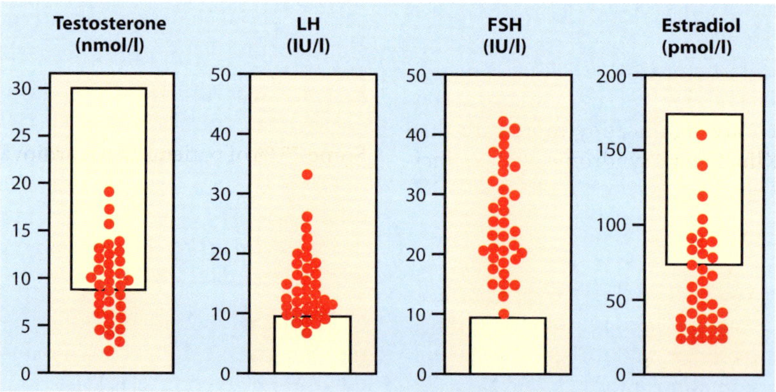

Figure 2.10: Hormone analysis of 35 men with Klinefelter's syndrome (47,XXY). The yellow areas give the reference ranges. Unlike luteinizing hormone (LH), follicle stimulation hormone (FSH) is almost always increased and the testosterone levels of many men are significantly reduced. In contrast to the literature, patients do not show increased serum estradiol levels.

significantly less costly and stressful alternative to attain fatherhood is to adopt.

The androgen deficiency must be compensated as soon as there are clinical signs of endocrine testicular insufficiency, or when laboratory findings confirm this condition. If total testosterone is still normal but LH is already increased, substitution should not be delayed, as the increased LH acts as an '*in vivo* bioindicator' of androgen deficiency already present.

Androgen treatment should also start early in the adolescent, as this significantly promotes their psychosocial development[280]. Testosterone therapy administered at the appropriate time may possibly prevent the development of gynecomastia and reduce the risk of breast cancer. Psychological counselling of the patient and their family should be given if possible before problems develop.

If the gynecomastia is cosmetically disturbing, mastectomy by an experienced breast surgeon is indicated. Although a prophylactic mastectomy is not usually necessary, the patients should examine their breasts regularly and undergo mammography where necessary.

2.2.3.5. XX-Man

Definition and incidence

Despite having the female genotype of 46,XX, affected patients possess neither internal nor external female sexual organs and are phenotypically purely male. In approximately 75 % of these patients, a translocation has taken place between the X- and Y- chromosomes in the spermatogenesis of the father, in which the gene for testicular structure (sex-determining region Y, SRY) was transferred[108]. XX-men are significantly more rare than patients with Klinefelter's syndrome, with an incidence of 1 : 20 000.

The clinical picture is very similar to that of Klinefelter's syndrome (small, firm testicles, gynecomastia, azoospermia); however, XX-men have normal body proportions and tend to be somewhat smaller than the population average. XX-men more often suffer from hypospadia. As in Klinefelter's syndrome, serum levels of testosterone are normal or reduced, estrogen and gonadotropins increased and the semen reveals azoospermia. Karyotyping is the only way to differentiate between the two syndromes. Treatment is the same as for Klinefelter's syndrome.

2.2.3.6. Noonan syndrome

Definition

Noonan syndrome covers a heterogeneous group of men who display the clinical and morphological symptoms of the female Turner's syndrome, combined with reduced testicular function. Only 5 % of patients have the genotype 45,X0, usually in the form of mosaics (45,X0/46,XY). In this 'true' Noonan syndrome, as in the XX-man, a transfer of genetic material has taken place from the Y-chromosome to other chromosomes with subsequent loss of the Y-chromosome[96].

Nevertheless, over 95 % of patients have a normal karyotype (46,XY), and are therefore not regarded as carriers of Turner's syndrome (45,X0) in women. These men with normal karyotype have other syndromes with a similar clinical picture (e.g. William's syndrome).

Clinical picture

The main symptoms are similar to those of Turner's syndrome:

- Short neck with pterygium colli
- Proportional microsomia
- Ptosis
- Pectus excavatum with supernumery nipples
- Epicanthus
- Micrognathia
- Abnormalities of the external ear
- Cubitus valgus
- Undescended testicle (in 70 % of patients)

Some 75 % of patients have cardiovascular malformations and heart defects which may reduce life expectancy and hence lead to clinical symptoms. Approximately one-half of patients are mentally retarded. The majority are infertile as there are no gametes in the testes. Testosterone is often reduced and gonadotropins increased, indicative of primary hypogonadism.

Therapy

Correction of the cardiovascular abnormalities is paramount. If the testicles are undescended the testes should be placed into the scrotum because of the increased risk of degeneration. The infertility cannot be treated. Testosterone treatment is indicated if there is androgen deficiency.

2.2.3.7. XYY-syndrome

Definition and incidence

This aberration in the number of chromosomes involves an extra Y-chromosome, and, therefore, the karyotype is 47,XYY. The reason for the aberration in the number of chromosomes is a disorder of chromosome distribution in the father's meiosis. The incidence is approximately 1 in 1000 newborn males.

Clinical picture

Except for their excessive height, most patients do not show clinical signs, and therefore the diagnosis is often made by chance. Testicular function is often not or only slightly impaired. The majority of patients are sufficiently androgenized and fertile. However, spermatogenesis may be disturbed, meaning infertility which cannot be treated. Some patients suffer from testosterone deficiency which requires substitution. It was previously thought that men with the karyotype 47,XYY were very aggressive, but this has now been disproofed[387]. However, cognitive abilities can be reduced. A diagnosis is made by determining the extra Y-chromosome under a fluorescent microscope, and karyotyping.

2.2.3.7.1. Male pseudohermaphroditism

The term male pseudohermaphroditism refers to various disorders of sexual differentiation which, despite the male genotype (46,XY), lead to incomplete or totally absent masculinization. In addition to genetic disorders of testicular structure and differentiation, any impairment of production or effect of the testicular hormones essential for sexual differentiation (testosterone, DHT and anti-Müllerian hormone (AMH)) can cause male pseudohermaphroditism (☞ Table 2.12)[192]. In rare cases, male pseudohermaphroditism is associated with congenital malformations in the form of syndromes[194]. Therefore, male pseudohermaphroditism is a symptom, not a diagnosis. A whole section is devoted to target organ defects (androgen resistance, 5α-reductase deficiency and oviduct persistence) in Chapter 2.2.4.

Genetic defects
• partial or complete loss of SRY • 46,XY gonadal dysgenesis with variants • 46,XY testicular agenesis • associated with congenital syndromes (e.g. Drash syndrome)
Testicular defects
• Leydig cell agenesis and hypoplasia (LH receptor defect) • Enzyme defects of testosterone biosynthesis - 20,22-desmolase deficiency (cytochrome P450scc deficiency, CAH type VI) - 3β-hydroxysteroid dehydrogenase deficiency (CAH type IV) - 17α-hydroxylase deficiency (cytochrome P450c17 deficiency, CAH type V) - 17,20-desmolase deficiency (cytochrome P450c17 deficiency) • 17β-Hydroxysteroid dehydrogenase deficiency • Anti-Müllerian hormone deficiency (oviduct persistence)
Target organ defects
• 5α-Reductase deficiency • Androgen receptor defects (androgen-resistance syndrome) - testicular feminization (complete, incomplete) - Reifenstein's syndrome - infertile men - undervirilized men • Defect of anti-Müllerian hormone receptor

Table 2.12: Classification of male pseudohermaphroditism[192]. SRY = sex determining region Y, gene of testicular structure; CAH = congenital adrenal hyperplasia.

2.2.3.7.2. Gonadal dysgenesis

The term gonadal dysgenesis (defective gonadal development) covers a number of disorders of gonadal differentiation which are caused by aberrations in the number and structure of chromosomes. A distinction is made between 46,XX and

46,XY gonadal dysgenesis and variants with excess gonosomes, depending on karyotype. If karyotype 46,XY is present, the disorder is described as simple gonadal dysgenesis (Swyer's syndrome)[217]. Despite the male karyotype 46,XY, the fetal gonadal projection does not develop into testes in these patients. In approximately 10 % of patients, defects are present in the gene on the Y-chromosome responsible for differentiation of the testes (sex-determining region Y, SRY). In addition, more recent studies indicate that mutations in the DAX-1 and SF1 genes may also cause gonadal dysgenesis and male pseudohermaphroditism.

Clinical picture, diagnosis, therapy

As there is no, or hardly any, endocrine activity in the gonads, a sexual differentiation disorder occurs. Accordingly there is usually complete feminization, and affected women consult the doctor initially for primary amenorrhea. The intersexuality may be noticed in the newborn by hypertrophy of the clitoris or incomplete labioscrotal fusion due to remaining activity of testicular endocrine function. As the Fallopian tube, uterus and vagina are usually laid down, these children are classified as girls. The gonads can be identified by means of sonography or MRI as a 'gonadal streak' in the region of the genital ridge or in the position of the ovaries, and should be removed before puberty owing to the high risk of malignant degeneration (gonadal blastoma), and also because these gonads can lead to undesirable masculinization during puberty as a result of the increased testosterone synthesis. Since there is also no estrogen synthesis, it is necessary to substitute estradiol to form secondary female characteristics and prevent eunuchoid body proportions.

Largely masculinized infants are noticeable by hypospadia and cryptorchidism (undescended testicles). Puberty does not occur, and therefore this must be initiated using testosterone. The undescended gonads should be removed in boys too, because of their tendency to degenerate.

Laboratory diagnosis, carried out at the latest by the time puberty fails to occur, will reveal the sharp increase in gonadotropins combined with the low concentration of sex steroids, confirming primary hypogonadism. Further confirmation is provided by the absence of testicles and the 'gonadal streak'. If there is any doubt about the diagnosis, this can be confirmed prior to gonadectomy by laparoscopic biopsy. While the infertility cannot be corrected, the absence of sex steroids can be compensated by substitution for the rest of the patient's life.

2.2.3.8. Testicular defects

Leydig cell aplasia or hypoplasia

Homozygotic mutations in the LH receptor gene (chromosome 2p21) cause Leydig cell aplasia or hypoplasia, which is very rare. Several families have been identified with this defect to date[358,394]. Although LH is present, it is unable to act because of the inactivating mutation of the receptor, so the Leydig cells do not develop properly and no testosterone is formed. Therefore, although there are testes present, there will be no increase in testosterone in the hCG test. Newborns affected with this condition display feminization or intersexual orientation of the external sexual organs. As anti-Müllerian hormone continues to be secreted by the intact Sertoli cells, oviduct derivatives are absent. The testicles usually lie within the abdomen.

Enzyme defects of testosterone biosynthesis

Biosynthesis of testosterone from cholesterol is catalyzed in the Leydig cells by five different enzymes (☞ Figure 1.4). Any of these enzymes may be absent or its activity may be reduced. Defects of testosterone biosynthesis are extremely rare and are due to inherited autosomal-recessive gene defects (☞ Table 2.13).

As steroid biosynthesis in the testes and adrenal glands partially requires the same genes, symptoms of hypogonadism and corticoadrenal insufficiency can occur simultaneously[300].

A complete absence or sharp reduction in the activity of one of the enzymes causes non-production or inadequate production of testosterone and excessive enrichment of the substrate before the enzyme block. Accordingly, the Wolffian duct does not become differentiated into the male sex organs (epididymis, vas deferens, seminal vesicle), and the external genitalia are not masculinized.

The external genitalia are feminized to a varying extent, depending on the remaining activity of the defective enzyme (☞ Table 2.13). If there is a complete enzyme block, female genitalia are usually formed. These are unambiguously female and of-

Absent enzyme						
Parameter	20,22-desmolase	3β-HSD	17α-OH	17,20-desmolase	17β-HSD	5α-Reductase
Enzyme protein	P450scc	no P450	P450c17	P450c17	no P450	no P450
Gene location	15q23-q24	1p11-p13	10q24-q25	10q24-q25	17q11-q12	2
Incidence	< 50 patients	< 100 patients	< 1 : 100 000	< 50 patients	1 : 100 000	< 200 patients
CAH	type VI	type IV	type V	-	-	-
DHT	absent	absent	absent	absent or reduced	absent or reduced	absent
Testosterone	absent	absent	absent	absent or reduced	absent or reduced	normal
Cortisol	absent	absent	absent	normal	normal	normal
Aldosterone	absent	absent	absent or reduced	normal	normal	normal
Main products of androgen synthesis (increased steroids)	-	DHEA, 17α-OH-pregnenolone	DOC, corticosterone, progesterone	17α-OH-pregnenolone, 17α-OH-progesterone	androstenedione, estrone	testosterone
Pituitary hormone	LH ↑, FSH ↑, ACTH ↑	LH ↑, FSH ↑, ACTH ↑	LH ↑, FSH ↑, ACTH ↑	LH ↑, FSH ↑	LH ↑, FSH ↑	normal
External genitalia	female	male/intersexual	female/intersexual	female/male/intersexual	female	female/intersexual
Wolffian duct	absent	normal	absent	rudimentary	hypoplasic	normal
Müllerian duct	absent	absent	absent	absent	absent	absent
Other	high mortality	high mortality/alkalosis	hypertension, hypokalemia	virilization at puberty	virilization at puberty	virilization at puberty
Gonads	testes	testes	testes	testes	testes	testes
Karyotype	46,XY	46,XY	46,XY	46,XY	46,XY	46,XY

Table 2.13: Enzyme defects of testosterone biosynthesis. Gene location indicates the chromosome which carried the gene for the enzyme. Modified from reference [279].
CAH = congenital adrenal hyperplasia; 3β-HSD = 3β-hydroxysteroid dehydrogenase; 17α-OH = 17α-hydroxylase; 17β-HSD = 17β-hydroxysteroid dehydrogenase; P450 = cytochrome P450; DHT = 5α-dihydrotestosterone; DHEA = dehydroepiandrosterone; DOC = deoxycorticosterone; LH = luteinizing hormone; FSH = follicle stimulating hormone; ACTH = adrenocorticotropic hormone.

ten do not arouse suspicion of intersexuality. However, the vagina ends in a blind pouch, as the uterus, Fallopian tubes and ovaries are always missing owing to the continued secretion of anti-Müllerian hormone. The testes do not fully descend and can be found inside the abdomen, in the groin or, more rarely, in the labia majora. Occasionally diagnosis is made before puberty if a herniotomy is performed. The enzyme defect is normally discovered when investigating primary amenorrhea.

However, if enzyme activity is only slightly reduced, predominantly male genitals are formed, with hypospadia, incomplete labioscrotal fusion and malpositioning of the testes. There is huge scope for variation, and therefore all transitional forms are possible.

2.2.3.8.1. Differential diagnosis and therapy of male pseudohermaphroditism

If a child is born with intersexual genitals, its sex should be determined rapidly, to pacify the parents and so it can be named. This should be done at a specialist endocrinology clinic which has the appropriate diagnostic instruments.

Examination of the newborn, supported by sonography and MRI, can help identify the structure of the uterus and Fallopian tubes and locate the gonads (☞ Figure 2.11). If female fallopian tubes are found, male pseudohermaphroditism is very unlikely.

A gonad outside the inguinal ring is almost always a testicle or an ovotestis, as an ovary practically never descends below the external inguinal ring.

Exclusion of defects of adrenal steroidogenesis (congenital adrenal hyperplasia) plays a central role in the diagnosis. If determination of 17-hydroxyprogesterone, progesterone and dehydroepiandrosterone, if necessary after stimulation with ACTH, does not show any evidence of an enzyme defect, karyotyping is carried out. This often leads to a diagnosis of Klinefelter's syndrome (47,XXY) or variants of the condition, X-chromatin-negative gonadal dysgenesis (46,XY/45,X0) or true hermaphroditism (46,XX/46,XY), which is confirmed by biopsy of the gonads and discovery of ovotestes. A normal male karyotype 46,XY necessitates investigation of testicular testosterone production with the hCG test. If the base testosterone concentration doubles after 72 h, this is normal. However, an insufficient increase in testosterone after 120 h indicates an enzyme defect of testosterone biosynthesis, Leydig cell aplasia or anorchidism. It is possible to distinguish between enzyme defects by determining characteristic patterns of different steroids in the serum.

The determination of basal 5α-dihydrotestosterone and 5α-dihydrotestosterone in the hCG test with calculation of the testosterone/5α-dihydrotestosterone ratio monitors the activity of 5α-reductase. Inadequate production of testosterone and 5α-dihydrotestosterone in a chromosomally male newborn (46,XY) who is not sufficiently masculinized indicates an androgen receptor defect, which is verified by detecting a mutation in the androgen receptor gene. If the urogenital sinus and external genitalia are that of a normal male, but the testes have not descended and the penis has not grown, hypothalamic and hypophyseal functional disorders (secondary hypogonadism) should be considered.

In the newborn who has intersexual external genitalia, the decision about establishing gender and any surgical measures is driven by the later functioning of the genital organs. Male genotype and testes play a subordinate role in reaching a decision.

Patients whose genitalia are predominantly feminized can undergo female reconstruction of the perineum and phallus. The testes are removed to prevent any malignant degeneration and possible virilization during puberty. During puberty, the patient is then given long-term estrogen substitution.

In patients with male genitalia on the outside, correction of the hypospadia is essential, possibly after stimulating penile growth with androgens. In oviduct persistence, the uterus and Fallopian tubes are removed, leaving the spermatic duct which runs into the ligamenta teretia uteri. The testes are placed in the scrotum. Apart from oviduct persistence, most forms of pseudohermaphroditism require testosterone substitution, which is started at the time puberty is expected and continued permanently thereafter.

2.2.4. Target organ-related causes of male hypogonadism

The term target organ defects covers disorders in which the testicular hormones testosterone and anti-Müllerian hormone (AMH) and/or the testosterone metabolites DHT and estradiol do not function, despite testis function being essentially intact. This is due either to defects in the corresponding receptors, or to a lack of activity by the enzyme 5α-reductase or aromatase, which are responsible for converting testosterone to DHT or estradiol.

2.2.4.1. Androgen resistance (androgen receptor defects)

> Androgen resistance is due to a defect of the androgen receptor in the androgen target organs, so that androgens are reduced or have no effect at all.

Androgen resistance is caused by sporadic or familial mutations of the androgen receptor gene. With a positive family history there is an X-chromosomal recessive inheritance, as the androgen receptor gene lies on the X-chromosome (Xp11-12). To date, almost 250 different gene mutations have been identified, 200 in exons B-H, which carry the steroid and DNA-binding domains, and only 23 in exon A (☞ Figure 1.7)[59]. Nevertheless, mutations in exon A almost always lead to complete loss of

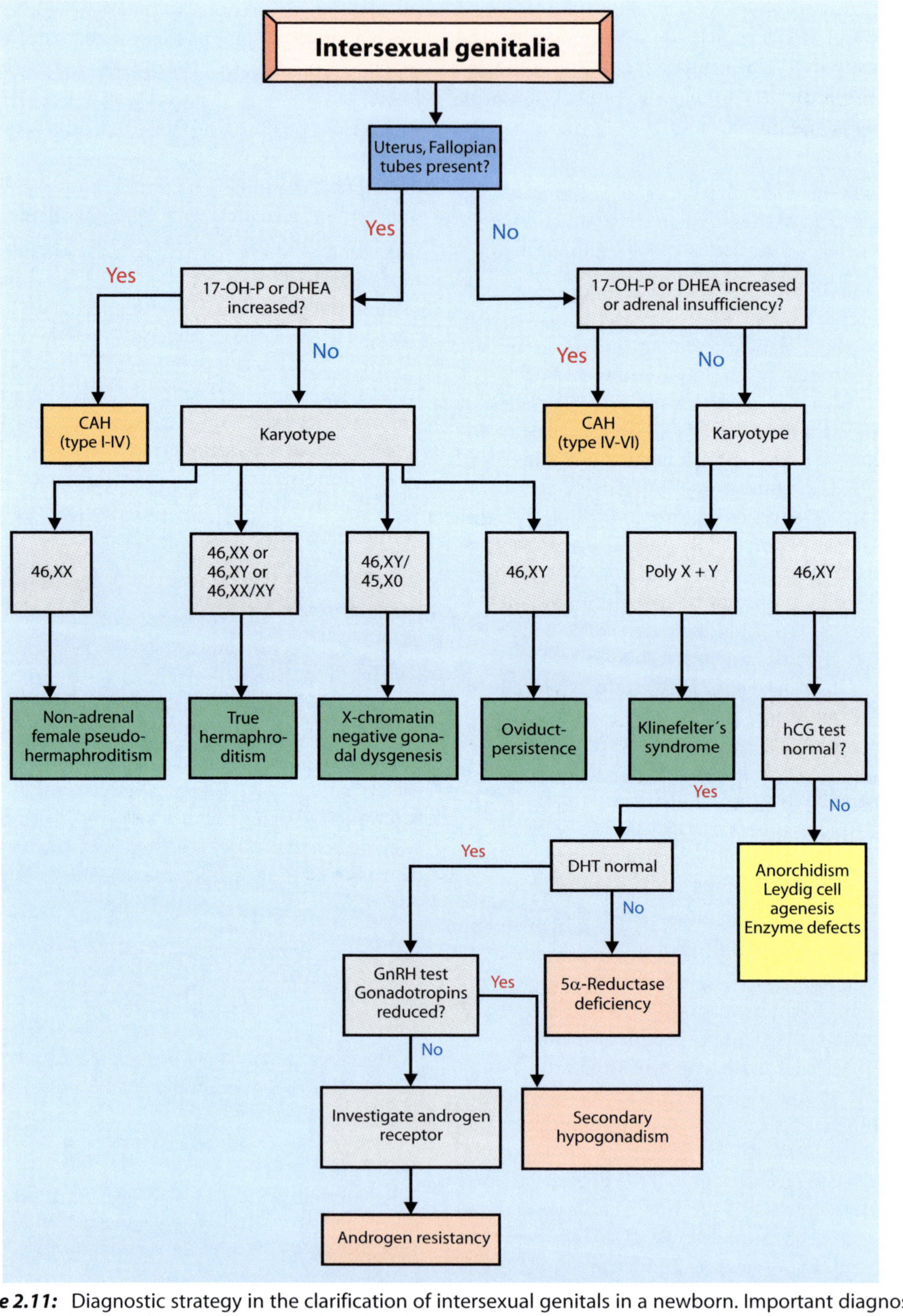

Figure 2.11: Diagnostic strategy in the clarification of intersexual genitals in a newborn. Important diagnostic steps include karyotyping and hormone determination to detect/exclude congenital adrenal hyperplasia (CAH). 17-OH-P = 17-hydroxyprogesterone; DHEA = dehydroepiandrosterone; NN-Insuff. = suprarenal insufficiency; DHT = 5α-dihydrotestosterone; hCG = human chorionic gonadotropin; GnRH = gonadotropin releasing hormone.

function of the androgen receptor, whereas mutations in other exons may cause only slight fluctuations in function. A database containing all mutations of the androgen receptor is available on the internet (www.mcgill.ca/androgendb).

However, an increase in polyglutamine sequences (CAG) to over 38 in exon A, in addition to causing slight androgen resistance, also surprisingly leads to a rare neurological disease which leads to progressive spinal and bulbar muscular atrophy (Kennedy's syndrome)[224,428].

In Kennedy's syndrome, a progressive degeneration of motor neurons in the spine manifests in early to mid-adulthood. The mechanism and relationship to androgen receptors is not at all clear. In addition, those affected display slight but progressive androgen resistance, which frequently manifests as gynecomastia, a reduced androgen effect (progressive loss of libido and sexual potency) and increasing infertility. However, many patients have already fathered children, and therefore the pathologically high number of CAG repeats in the androgen receptor is inherited.

As patients with mutations of the androgen receptor in other exons or total loss of androgen receptors do not suffer from this condition, the N-terminal section (exon A) of the androgen receptor appears to exert as yet unknown functions in motor neurons. Research in this area might yield very interesting findings.

In addition, the number of CAG repeat sequences in exon A is associated with the development of carcinoma of the prostate. Fewer than 12 repeats are thought to increase the risk of developing carcinoma.

Clinical picture

In androgen resistance, functional disturbance of the androgen receptor can vary in severity (☞ Table 2.14). This results in a pathophysiological continuum of restricted function, which is reflected in the wide-ranging clinical phenotype of affected patients, which shows fluid transitions. Complete testicular feminization represents the extreme variant, which exhibits continuous transitions through partial testicular feminization, Reifenstein's syndrome and infertile men to insufficiently virilized men. Therefore, forms of testicular feminization lead to male pseudohermaphroditism.

The incidence of androgen resistance syndrome is not accurately known. It is thought that 1 in 20 000-64 000 newborns has an androgen receptor defect.

Testicular feminization

Complete testicular feminization is due to complete loss of function of the androgen receptor[133]. The external genitalia are unambiguously female, and therefore patients affected with this condition are seldom recognized before puberty. The internal genitalia are missing, as testicular secretion of AMH, which was not impaired, prevented the development of a uterus and Fallopian tubes, and stimulation of the epididymis, seminal vesicle, duct and prostate by androgens did not occur owing to the androgen receptor defect. The vagina ends in a blind pouch and is short, but usually adequate for sexual intercourse. The testes lie within the abdomen, in the groin or in the labia majora. Occasionally diagnosis is made before puberty if a herniotomy is performed. Approximately 2 % of all girls with inguinal hernias have an androgen receptor defect. Normal or hyperplastic Leydig cells are found on histology. Spermatogenesis is incomplete.

Puberty with thelarche, intermittent growth and body fat distribution, as well as psychological development, is as for normal females; however, menstruation and secondary hair are completely absent (☞ Figure 2.12). Testicular feminization is the third most common cause of primary amenorrhea, and approximately 10 % of all women with primary amenorrhea have testicular feminization. Primary amenorrhea or inability to conceive usually causes the patient to consult a doctor.

If sonography shows ovaries, Fallopian tubes and the uterus to be missing, chromosome analysis should be instigated. This will reveal a 46,XY karyotype. Investigating hormones in the serum will show up androgen resistance. With lack of negative feedback to the hypothalamus and hypophysis, serum LH is increased and androgen production increases above its normal level. As a result, serum levels of testosterone and often DHT and estradiol too are increased. Serum FSH is usually normal, despite defective spermatogenesis. The suspicion of testicular feminization is confirmed by biomolecular investigation of the androgen receptor gene.

	Complete testicular feminization	Partial testicular feminization	Reifenstein's syndrome	Infertile men	Subvirilized men	5α-Reductase deficiency
Karyotype	46,XY	46,XY	46,XY	46,XY	46,XY	46,XY
Gonads	testes	testes	testes	testes	testes	testes
Fallopian tubes/uterus	absent	absent	absent	absent	absent	absent
Epididymis/ vas deferens/ seminal ves-icle	absent	hypoplastic	male or hypoplastic	normal male	normal male	normal male
Prostate	absent	absent	hypoplastic	normal	normal	normal
External genitalia	female	female	male	male	male	female
Fertility	absent	absent	absent or very reduced	absent or very reduced	absent or reduced	absent
Develop-ment of breasts	female	female	gyneco-mastia	male or gyneco-mastia	male or gyneco-mastia	male
Secondary hair	absent	female	sparse male	normal	spare male	sparse male
LH	↑	↑	↑	normal or ↑	normal or ↑	normal
Testosterone	↑	↑	normal or ↑	normal or ↑	normal or ↑	normal
5α-Dihydro-testosterone	↑→	↑	normal or ↑	mostly normal	mostly nor-mal	absent
17β-estradiol	↑	↑	↑	mostly nor-mal	mostly nor-mal	normal
FSH	normal or ↑	normal or ↑	normal or ↑	normal or ↑	normal or ↑	normal or ↑

Table 2.14: Clinical symptoms of varying degrees of androgen resistance (androgen receptor defects). To make a differential diagnosis, 5α-reductase deficiency is also included. LH = luteinizing hormone; FSH = follicle stimulating hormone.

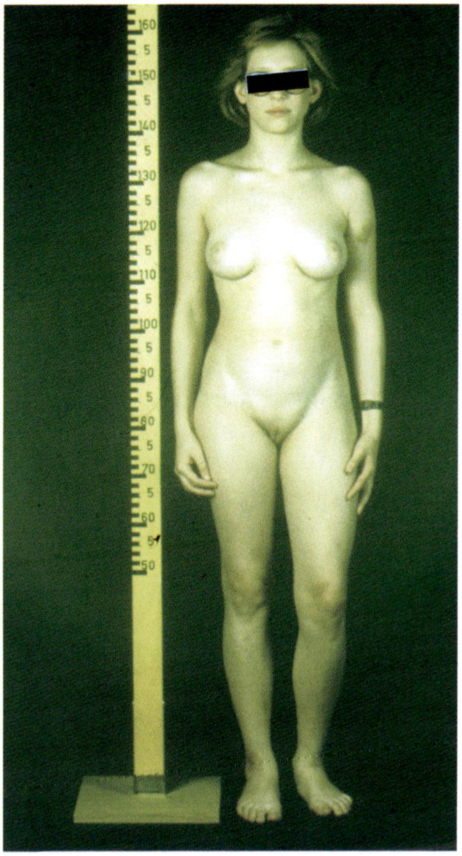

Figure 2.12: A 20-year-old woman with complete testicular feminization, primary amenorrhea, karyotype 46,XY, missing fallopian tubes and uterus. The testes lie in the inguinal canal and can be seen as bulges on the lateral corner of the mons pubis. Figure courtesy of Dr. M. Bals-Pratsch, Regensburg.

Approximately 10 % of female patients have an incomplete form of testicular feminization. A slight residual function from the androgen receptor causes slight masculinization of the internal and external genitalia of these patients. The clitoris is enlarged and the labia majora are not completely fused. Whereas the derivatives of the Müllerian duct are completely missing, the derivatives of the Wolffian duct are partly formed. There is normal thelarche at puberty. Unlike complete testicular feminization, the clitoris increases in size and secondary hair starts to grow. As the testes have a tendency to become malignant, these should be removed[287]. In patients with the incomplete form of the condition, this should, if possible, be performed before puberty to prevent the slight virilization that occurs around this time. Patients with complete testicular feminization should also have their testes removed before puberty if the testes are easily accessible within the groin or labia. Otherwise the testes should not be removed until after puberty, as the endogenous production of estrogen promotes development of puberty and virilization does not occur. After gonadectomy, all patients are given estrogen therapy which, in prepubescent patients, starts at the time of onset of puberty. It is better to withhold the diagnosis of testicular feminization from the patient, as the knowledge that they have a male genotype and testes or male gonads can trigger a severe psychological crisis, affecting how a woman sees herself and causing personal breakdown or breakdown of a marriage. It is best to explain the amenorrhea and untreatable infertility in terms of defective gonads which have to be removed because they do not function and have a tendency to degenerate. Women with testicular feminization have normally fully identified with their phenotypical gender by the time of diagnosis. They can have a completely normal marriage and become good adoptive parents.

Reifenstein's syndrome

This group of patients suffers from the next weakest variant of androgen resistance. Unlike testicular feminization patients, Reifenstein patients are almost always identifiable as males at birth. However, the external genitalia are usually insufficiently masculinized; this can range from incomplete labioscrotal fusion with formation of a pseudo vagina, through perineoscrotal hypospadia, to slight penile hypospadia[11]. The testes are often not completely descended. More rarely, there is only gynecomastia and azoospermia, which become apparent only after puberty. The derivatives of the Wolffian duct are normal or hypoplastic. Different phenotypes of the same gene defect may manifest within one family[239]. Increasing virilization sets in at puberty, which is often delayed, and physical characteristics are predominantly male. Pubic and armpit hair are normal, but remaining body and facial hair is sparse. The voice may not break. All patients develop gynecomastia, which can be disturbing (☞ Figure 2.13).

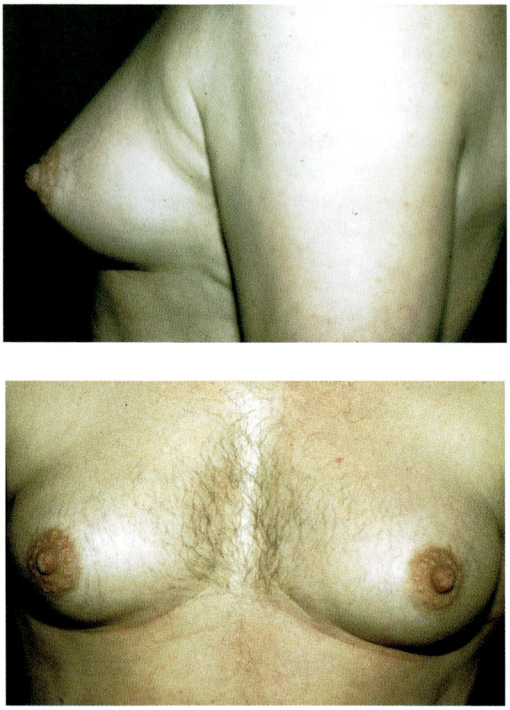

Figure 2.13: Pronounced gynecomastia in a man with Reifenstein's syndrome. For cosmetic reasons, a bilateral mastectomy was carried out by an experienced breast surgeon.

However, there is no evidence of an increased incidence of breast cancer. The testes often remain smaller than normal as a result of atrophy of the germinal epithelium. The histological findings range from complete hyalinization of the seminiferous tubules through germinal cell aplasia to almost completely normal findings with arrest of spermatogenesis at the primary spermatocyte stage[297]. A spermiogram performed on patients with Reifenstein's syndrome usually shows azoospermia. The Leydig cells appear normal or hyperplastic as a consequence of overstimulation by LH, which, because of the androgen receptor defect, reacts inadequately to the feedback of testosterone. Therefore, the amplitude and frequency of LH pulses are increased. In the GnRH test there is an excessive increase in both of the gonadotropins. Serum levels of LH and estradiol, and usually testosterone too, are increased. FSH levels may be normal or increased. The clinical finding of gynecomastia combined with azoospermia or severe oligoospermia, especially when associated with signs of defective sexual differentiation (e.g. hypospadia, undescended testicles) should lead to consideration of a diagnosis of Reifenstein's syndrome. Increased or high-normal LH, estradiol and testosterone concentrations strengthen the suspicion, which can then be confirmed by biomolecular determination of an androgen receptor defect. There are different degrees of this syndrome, ranging to complete infertility[259].

Infertile and insufficiently virilized men with androgen resistance

Infertile-man syndrome is a mild form of androgen resistance. The only symptom is defective spermatogenesis, which causes infertility. The external genitalia are unambigously male and the testes have descended. Typically, the patients are conspicuous by azoospermia or severe oligospermia, combined with increased LH and testosterone concentrations. However, these criteria are not sufficient to distinguish them from men with infertility caused by other factors. To secure a diagnosis, it is necessary to investigate the androgen receptor gene using biomolecular techniques. Nevertheless, androgen receptor defects are rarely the cause of idiopathic infertility[6,53,276].

Insufficiently virilized men with gynecomastia, small penis, sparse facial and body hair, normal urethra and normal sperm density are a further variant of mild androgen resistance. Some of these men are fertile[135].

In Reifenstein's syndrome and insufficiently virilized men, high-dose testosterone therapy can be attempted, to encourage virilization. Patients with mutations of the androgen receptor gene in the coding region for androgen binding appear to benefit more from this type of treatment than patients with mutations in the DNA-binding region[392]. It is not possible to make a general dose recommendation because of the lack of documented individual cases. Elevated gonadotropins and SHBG will gradually decrease, if the patient responds to treatment.

2.2.4.2. 5α-reductase deficiency

In the Dominican Republic, a large tribe was identified in which a number of boys with the male karyotype 46,XY had female external genitalia. Several families with this form of male pseudohermaphroditism have since been identified throughout the world. Testes are laid down, but uterus and Fallopian tubes are absent (☞ Table 2.11).

This disorder of sexual differentiation is caused by a mutation of the gene for the enzyme 5α-reductase type II on chromosome 2[388]. Testosterone functions as a prohormone for 5α-dihydrotestosterone (DHT), and is metabolized to this by the enzyme 5α-reductase. During the sexual differentiation phase, this enzyme is particularly active in the structures of the prostate and urogenital sinus, from which the external genitalia are formed. If there is no enzyme activity, these structures do not differentiate to the male phenotype, and therefore the external genitalia are female and there is no prostate. The testes lie in the labia majora, the inguinal canal or abdomen. However, the exclusively testosterone-dependent derivatives of the Wolffian duct - the epididymis, seminal vesicle and spermatic duct - are differentiated as a normal male, and end in a pseudo vagina. Depending on any residual activity of 5α-reductase, the concentration of DHT can rise during puberty and with increasing testosterone concentrations owing to the increased amount of substrate available, and masculinization can occur, with hypertrophy of the clitoris. Young boys previously brought up as girls occasionally change their sexual identity.

It is not clear why normal levels of testosterone are inadequate to induce male differentiation of the derivatives of the urogenital sinus, although they are adequate to differentiate the derivatives of the Wolffian duct. Maybe there are sufficiently high concentrations of androgen in the immediate vicinity of the Wolffian tract due to the paracrinal secretion of the Leydig cells, whereas the concentrations which reach the urogenital sinus by the endocrine route are not sufficient. If the activity of 5α-reductase is intact, this is possibly compensated by the higher affinity of DHT to the androgen receptor, compared with testosterone.

2.2.4.3. Oviduct persistence

To date, approximately 150 men have been described with normal karyotype (46,XY) and male external and internal genitalia who show rudimentary or fully formed derivatives of the Müllerian duct (uterus, Fallopian tubes). These patients have a disorder in the production (synthesis defect) or effect (receptor defect) of the anti-Müllerian hormone (AMH). Normally, the Sertoli cells in the testes produce AMH during the 9th-20th week of pregnancy. This causes atrophy of the Müllerian duct, which is also present in male fetuses. If there is no AMH, or it is ineffective owing to an AMH receptor defect, the Müllerian duct differentiates to Fallopian tubes and uterus, regardless of genetic and gonadal gender[231]. Both events, mutations of the gene for AMH (19p) and AMH receptor (12q), have been described in fewer than 70 patients with oviduct persistence.

2.2.4.4. Absence of estrogen effect in man

As testosterone functions as a prohormone for estrogens in man, target organ-related disorders include estrogen receptor defects and disorders of metabolism resulting from a deficiency of the enzyme aromatase must also be understood. To date, four men have been identified with a mutation in the gene for aromatase[268]. These men are unable to manufacture estradiol from the prohormone testosterone. In addition, one man has been discovered with an inactivating mutation in the estradiol-β receptor[365]. The absence of estrogen effect was particularly noticeable in the bones of all patients affected. The epiphyseal cartilage does not close; therefore, despite being over 20 years old, the patients are still growing and display eunuchoid body proportion with a height of 200 cm (= 6 feet 7 inches) and above. Moreover, they display markedly reduced mineralization of the trabecular and cortical bone, and therefore suffer from severe osteoporosis. In three patients with aromatase deficiency, the administration of estradiol led to closure of the epiphyseal cartilage and a sharp rise in bone mass. According to this, estradiol, and not testosterone as previously thought, is the crucial sex steroid for bone metabolism. Aromatase is present in osteoblasts and osteoclasts; therefore, the requirement for estrogens can be met by local production from testosterone. Apart from eunuchoid body proportions, all men were virilized normally, and their puberty was uneventful. Analysis of semen from the man with the estrogen receptor defect revealed asthenoteratozoospermia with normal sperm count. Some of the (few) men with inactivated estrogen receptor-α or aromatase deficiency have functional disorders of the testes[268]. This is confirmed by findings from animal studies. The Sertoli and Leydig cells, the epididymis and efferent testicular ducts express estrogen receptors, as do the gametes themselves. Mice with inactivated estrogen receptors are infertile as a result of a functional disorder of the epididymis and rete testis. In addition, these mice are completely disinterested in sex. Hence, estrogens are obviously an important stimulus of male sexuality. In adult humans too, estrogens appear to stimulate libido and sexual activity. Treating one of the men deficient in aromatase with estrogens increased libido, frequency of sexual fantasies, masturbation and sexual intercourse. Hormone analysis showed significantly increased gonadotropin levels, which emphasizes the significance of estradiol to negative feedback. Whereas the men with aromatase deficiency had increased androgen levels and no estradiol, those with estrogen receptor defects had normal levels of androgens and increased estradiol in their serum.

Moreover, the men had metabolic disorders with obesity, hyperinsulinemia (due to insulin resistance), hypertriglyceridemia and raised LDL cholesterol. Whereas, as expected, estrogen therapy caused no essential change in the man with the estrogen receptor defect, administration of estrogens to patients with aromatase deficiency caused the epiphyseal cartilage to close, significantly increased bone density and reduced the hyperinsulinemia and hypertriglyceridemia. Cholesterol was unaffected.

From these patients came the significant finding that, in man too, estradiol is the crucial factor in bone mineralization and closure of epiphyseal cartilage during puberty. This realization has enormous consequences for androgen treatment, since, with regard to bone metabolism, only androgens which can also be metabolized to estradiol are effective (☞ Chapter 6.).

2.2.5. Hypogonadism due to systemic diseases and exogenous toxins

Many systemic diseases impair the sensitive testicular and sexual function, sometimes considerably and irreversible. Therefore, hypogonadism as an early sign can contribute towards a premature diagnosis of the underlying condition. However, impairment of testicular function often goes unrecognized and untreated, although some underlying conditions would benefit from testosterone substitution.

Apart from particular diseases, practically any event that severely impairs health, e.g polytrauma, severe burns, sepsis, advanced autoimmune deficiency syndrome (AIDS), severe myocardial infarction, causes restricted testicular function. Numerous studies reveal a temporary deficiency of testosterone while the LH remains normal. This, like the so-called 'low-T3/T4-syndrome' of the thyroid gland, demonstrates the hypothalamic origin of hypogonadism. When health returns to normal, testosterone production normally recovers. It is not known whether substitution is beneficial in severely ill patients with testosterone deficiency.

2.2.5.1. Systemic (general) diseases

2.2.5.1.1. Liver cirrhosis

Alopecia of the abdomen, gynecomastia and testicular atrophy are some of the classical symptoms of liver cirrhosis in men. They are also the external signs of hypogonadism, which presents in 50-75 % of men with this condition. The etiology of liver cirrhosis is irrelevant for the development of hypogonadism, although hypogonadism occurs more frequently in alcohol-induced cirrhosis than in other forms of liver cirrhosis. Serum levels of testosterone are typically reduced, and those of estradiol and prolactin increased. Concentrations of the gonadotropins LH and FSH are usually normal[423].

The pathomechanism of hypogonadism appears to be initiated by the portocaval collateral circulations resulting from portal hypertension. Androgens are eliminated in the liver to a lesser extent through the bypass resulting from the collateral circulations, and therefore more androgens are available peripherally for conversion to estrogens. Through an increased negative feedback, estradiol causes suppression of the gonadotropins, which are usually normal despite reduced serum testostrone levels. In addition, estrogens stimulate production of SHBG in the liver, which is often raised in men with liver cirrhosis. As SHBG has a stronger affinity to testosterone than to estradiol, the estrogen-androgen balance shifts, and the amount of free testosterone is reduced. The increased concentration of estrogen is probably also responsible for hyperprolactinemia, which in turn favors the development of gynecomastia, together with the increased estrogens. In addition, there is Leydig cell insufficiency with reduced testosterone production, and a decreased reaction in the hCG test. The pathomechanism for this is unknown. It may be due to hyperestrogenemia.

Patients with liver cirrhosis have significantly reduced fertility. There is often oligoasthenoteratozoospermia. Histology of the testes shows peritubular fibrosis and atrophy of the germinal epithelium.

2.2.5.1.2. Enteropathies

Hardly any systematic studies have been carried out on the influence of inflammatory intestinal diseases. However, it has been observed that couples in which the man suffers from ulcerative colitis or Crohn's disease produce fewer pregnancies than couples in which the man is healthy. This provides evidence of a slight reduction in the fertility of men with enteropathies. Semen analysis corroborates the suspicion that fertility is impaired, as the findings are pathological in approximately 30-50 % of men with Crohn's disease. However, these investigations do not distinguish between specific effects of the disease and toxic effects of the medication. There is no clinical evidence of androgen deficiency in patients with Crohn's disease, and the majority of patients have normal serum testosterone and gonadotropin levels.

2.2.5.1.3. Kidney diseases

Disorders of testicular function occur in approximately 60 % of men with chronic renal insufficiency. These become noticeable as a general deterioration in sexual function long before the patient requires dialysis. Typically, total as well as free serum testosterone is significantly reduced and serum LH is increased. This is due to an insufficiency of the Leydig cells, which produce reduced basal testosterone and a low testosterone response to the hCG stimulation test. The persistent improvement in Leydig cell function following a kidney transplant indicates that azotemia *per se* is responsible for the restricted function of the Leydig cells.

Normal or even excessive stimulation of gonadotropins in the GnRH test, and the ability of LH to suppress exogenous testosterone, reveal hypophyseal function to be intact. Nevertheless, retention contributes to the increased concentration of gonadotropins in the serum, as 40 % of LH and 12 % of FSH is eliminated via the kidneys. Hemodialysis itself does not affect serum concentrations of testosterone, FSH and LH, as FSH and LH are not filtered and 90 % of testosterone in the serum is protein-bound so there is essentially no loss[194].

Whereas serum concentrations of SHBG and estradiol are usually normal, renal failure often increases hyperprolactinemia and leads to gynecomastia in many patients. Apart from reduced renal clearance of prolactin and frequent intake of prolactinogenic drugs, hyperprolactinemia is caused by a functional disorder of the lactotropic cells of the hypophysis, which respond less to inhibitory and stimulatory stimuli and therefore do not respond to this regulation. As hyperprolactinemia reverses after a kidney transplant, this indicates a causal connection with azotemia.

Men with terminal renal failure often have reduced fertility. The histological findings of testicular biopsies range from a quantitative reduction in gametes to almost complete germinal aplasia, where the stem cells are reduced to a lesser extent than in later stages of spermatogenesis. Semen analysis will therefore often show pronounced oligoasthenoteratozoospermia. Starting hemodialysis treatment will not prevent a further deterioration in semen findings and the steady increase in FSH which indicate progressive damage to the germinal epithelium. On the other hand, following kidney transplant, fertility may be restored as kidney function improves.

Erectile dysfunction in men with renal failure is caused not only by endocrine but also by many other factors. Neurogenic disorders resulting from uremia or diabetes mellitus, as well as vascular factors, contribute towards the potency disorder. In patients who have had a kidney transplant, the blood supply to the penis may be reduced, depending on closure of the vessels of the kidney at the internal iliac artery. In addition, many dialysis patients are taking a lot of medication, which is capable of inducing or intensifying a potency disorder.

Whereas there is evidence accumulated over a long period of time that testosterone therapy has a favorable effect on renal anemia, there is a lack of information on its effect on erectile dysfunction and loss of libido. As potency disorders can be caused by a variety of factors, it is anticipated that only a proportion of men with renal failure will improve if the testosterone deficiency is compensated. Nevertheless, testosterone deficiency should be treated, as normalizing testosterone levels are expected to have a favorable effect on renal osteopathy and to increase erythropoiesis so that the requirement for erythropoietin, which is an expensive treatment for renal anemia, will diminish. However, there are no studies available on this subject. Testosterone therapy causes a slight increase in serum creatinine in some patients, which is a consequence of the desired anabolic effect. As intramuscular administration of testosterone is problematical because dialysis patients receive regular heparin, testosterone therapy can be given in the form of gel.

2.2.5.1.4. Diabetes mellitus

There are no reports of diabetes mellitus having an adverse effect on testicular function. The prevalence of erectile dysfunction is approximately 6 % in young people with diabetes (20-24 years of age), increasing gradually with age. Some 50 % of older diabetics (55-59 years of age) complain of impotence. This correlates with the appearance of secondary complications (nephropathy, neuropathy and angiopathy)[252]. This indicates a secondary cause of impotence as a result of neuropathy and micro- and macroangiopathy. In addition, diabetics can also suffer from erectile dysfunction for other reasons, which is why appropriate investigations should be carried out to exclude psychogenic and endocrine causes[277].

Sperm production is usually not disturbed in diabetics. Nevertheless, many diabetics suffer from infertility resulting from an ejaculation disorder, which is a consequence of the neuropathy. If there is azoospermia or aspermia, to exclude retrograde ejaculation, urine obtained following orgasm is investigated. If sperm are found in the urine, there is a co-ordination disorder of the muscles of the pelvic floor and neck of the bladder, which can occasionally be corrected using parasympatholytics or α-sympathomimetics. If this is unsuccessful, an attempt can be made to inseminate sperm obtained from the urine.

Accordingly, therapy to prevent infertility and erectile dysfunction is centered around maintaining blood sugar levels as near to normal as possible. Intravenous administration of α-lipoic acid may have a favorable effect on peripheral neuropathy. There are no data available on erectile dysfunction. However, phosphodiesterase-5 inhibitors have proven to be very effective in diabetics with erectile dysfunction. If there is testosterone deficiency (although this is rare), this should be compensated. Testosterone substitution is not contraindicated in diabetes mellitus. In fact, recent data indicates a beneficial effect of testosterone therapy on insulin resistance.

2.2.5.1.5. Hemochromatosis

Approximately 60 % of patients with hemochromatosis also display symptoms of hypogonadism. The hypothalamus, hypophysis and testes are disturbed as a result of increased iron deposition. Hence, testosterone, LH and FSH are reduced to a variable extent. GnRH and hCG stimulation tests therefore often trigger a weak reaction. Prepubescent patients may experience delayed puberty. Blood-letting may improve the hypogonadism. Case reports indicate that testosterone therapy has a favorable effect on the course of the disease, as it increases erythropoiesis and makes it possible to bleed the patient more frequently.

2.2.5.1.6. Sickle cell anemia

Sickle cell anemia causes thromboses and multiple infarctions in the capillary system, which also affect the hypothalamus and testes. Approximately 30 % of all men with the homozygotic form of sickle cell anemia display symptoms of hypogonadism. This may be primary or secondary, or a combination of the two, depending on the location of the thromboses. Fertility is affected more often than testosterone production. Generalized substitution with testosterone cannot be recommended as the incidence of priapism is increased in this disease, and testosterone therapy causes an increase in hematocrit with increased risk of thrombosis.

2.2.5.1.7. Thyroid diseases

Disorders of thyroid function seldom impair fertility in men as they do in women. Thyroid releasing hormone (TRH)-mediated hyperprolactinemia in primary hypothyroidism may cause secondary hypogonadism. Occasionally, hyperthyroidism causes gynecomastia because of the increased hepatic production of SHBG. This results in increased estrogen activity. Decreased libido and sexual potency experienced by men may possibly be attributed to this. There have been isolated reports of pathological findings in the semen of men with hyperthyroidism which have returned to normal after the hyperthyroidism has been corrected.

2.2.5.1.8. Hypercortisolism

Supraphysiological concentrations of cortisol in the serum suppress production of LH in the hypophysis and cause secondary hypogonadism. This is demonstrated very clearly by the normalization of serum LH and serum testosterone after bilateral adrenalectomy (surgical or drug-induced) in patients with endogenous hypercortisolism. Exogenous cortisol also suppresses gonadotropins, which means that long-term therapy with corticosteroids can induce secondary hypogonadism. The catabolism induced by corticosteroids, as well as the osteoporosis and muscle atrophy that develop, is further intensified by testosterone deficiency. This is why testosterone substitution is indicated if testosterone levels are low in endogenous hypercortisolism that cannot be corrected, and when corticosteroids are being administered therapeutically.

Occasionally, men with Cushing's disease also develop gynecomastia. This leads to a sharp increase in the adrenal production of aromatized androgens such as androstenedione and dehydroepiandrosterone or, more rarely, estrogens.

2.2.5.1.9. Malignoma

The improved prognosis of many malignomas *quoad vitam* and the possibility of freezing sperm to lay down a store before the administration of cytostatics that can impair fertility has led to an increasing interest in the fertility of younger men with malignomas requiring treatment. Nevertheless, fertility studies almost exclusively cover patients with malignant lymphomas and gamete tumors. More than 50 % of patients display oligo- or asthenoteratozoospermia before therapy. The cause of the fertility disorder is unclear. FSH levels are often raised, indicating primary testicular damage. It is not clear whether cytokines, intermittent fever or the catabolism itself damages the epithelium. Testicular endocrinal function is rarely impaired.

2.2.5.1.10. Generalized infections

Human immunodeficiency virus (HIV)

HIV can affect every organ. The functioning of endocrine glands can also be disturbed by the retrovirus, opportunistic infections, neoplasia or toxic medication. A high percentage of men with HIV infection display symptoms of hypogonadism. Two-thirds of men with AIDS complain of a loss of libido, and one-third of impotence. The proportion of men with androgen deficiency increases as the disease advances, and is 50-70 % in AIDS. Men with asymptomatic seropositive HIV infection usually have normal serum testosterone levels. The behavior of serum LH is non-uniform, and may be high, normal or low. This reflects the differing origins of hypogonadism.

Autopsies carried out on a proportion of men with AIDS have revealed opportunistic infections of the testes due to toxoplasmosis, cytomegalovirus and tuberculosis. In addition, histology of the testes usually reveals a severe disorder of spermatogenesis with hyalinization of the tubuli, Sertoli-cell-only syndrome or arrest of spermatogenesis. These alterations are observed even where there is no simultaneous opportunistic infection. Therefore, in addition to testicular infection due to the retrovirus itself, the effects of fever and chemotherapy on the testes also have to be taken into account.

Secondary hypogonadism in patients with low serum testosterone and serum LH is predominantly the result of a hypothalamic disorder, as an adequate increase in LH is usually obtained in the GnRH test. It is caused by opportunistic infections (e.g. cerebral toxoplasmosis), as well as by weight loss and cachexia, which lead to hypogonadotropic hypogonadism in other serious diseases of catabolic metabolic conditions. In addition, pre-existing causes of hypogonadism should be considered in patients with HIV infection, such as drug abuse and urogenital infections.

Fertility, like endocrine function, appears to be reduced in men infected with HIV only at an advanced stage of the disease. Ejaculate parameters of asymptomatic seropositive men are normal, apart from the retrovirus being detected in the semen[87,100,222].

Tuberculosis

Tuberculosis affects the sexual organs, but this is rarely seen today. The prostate and epididymis are more likely to be affected. Tubercular orchitis is rare, but is included in the difficult differential diagnosis of testicular tumors. Occasionally, orchiectomy must be performed if non acid-proof rod-shaped bacilli can be detected in the urine or semen. Typical symptoms of tuberculosis of the testes and related sexual organs, in addition to a scrotal mass which is usually painless, include: hematospermia, sterile leukocyturia and increased leukocytes in the semen, which is often pathologically altered. Hydrocele can also be a sign of tuberculosis of the epididymis or testes. Androgen deficiency is rarely found in testicular tuberculosis, however, fertility is frequently impaired as there is usually oligoasthenoteratozoospermia[219].

Leprosy

Whereas leprosy is seldom observed in Europe, as many as 1 % of the population in some tropical countries are affected by this disease. Since leprosy causes a granulomatous infiltration of the testes, up to 60 % of men affected by the disease show symptoms of primary hypogonadism, depending on the type of leprosy and duration of the disease. Men with lepromatous leprosy are particularly affected. A series of autopsies carried out on such men revealed a 90 % infiltration of the testes. Some 5-10 % of men with borderline forms of the disease develop hypogonadism. Hypogonadism seldom occurs in tuberculoid leprosy. Patients display the typical clinical signs of androgen deficiency and are infertile. The gynecomastia occasionally observed is a consequence of the shift in the androgen-estrogen balance towards estrogens. The testes and epididymis may swell painfully as the disease progresses, and they may also become atrophied. Granulomatous alterations are seen on histology which may contain mycobacteria, which is why acid-proof rod-shaped bacilli are occasionally seen in the semen. There is usually oligozoospermia or azoospermia. Serum gonadotropins are usually increased, whereas serum testosterone is reduced[234].

2.2.5.1.11. Myotonic dystrophy

In this autosomal-dominant inherited syndrome with delayed muscle relaxation after contraction, dystrophy of the extremities and pharyngeal muscles, cataract, hyperacusis and frontotemporal hair loss, approximately 80 % of the men affected develop a progressive untreatable primary hypogonadism with testosterone deficiency and damage to the germinal epithelium. Testis volume and consistency are often reduced. Biopsy of the testes shows increasing hyalinization of the seminiferous tubules and vacuolation of the Sertoli cells. The Leydig cells are not normally pathological, and appear to be hyperplastic as a result of atrophy of the tubules. Nevertheless, 60 % of those affected with the disease have testosterone deficiency which causes loss of libido and loss of sexual potency. Gonadotropins are usually increased, the concentration of FSH usually exceeding that of LH, reflecting the damage done to the germinal epithelium. The pathomechanism behind primary hypogonadism is unknown.

If identified early, patients with myotonic dystrophy, or carriers of the disease, should be offered cryopreservation of their sperm. Substitution therapy can be given for testosterone deficiency. However, the resultant increase in muscle mass is not associated with increased muscle strength[212,273].

2.2.5.1.12. Transverse syndrome

Severe trauma to the spine will often cause exocrine testicular insufficiency. Whereas the production of testosterone often returns to normal spontaneously, spermatogenesis is usually permanently disturbed in paraplegics or quadraplegics. Disorders of thermoregulation and circulation are suspected causes. Isolated biopsies carried out to date showed a hyaline degeneration of the germinal epithelium.

The ability to have erections and ejaculate is dependent on the location of the transverse section. If the lesions are above the thoracolumbar junction (Th12-L1), psychogenic erection capacity with maintained reflex-induced erection, which is controlled by the sacral junction (L4-S2), will be absent. Reflex-induced erections usually cannot be maintained during coitus, as the psychogenic influences are missing.Tabes dorsalis (spinal atrophy) causes degeneration of the afferent somatosensitive fibers, and, therefore, as in sacral spinal lesions, reflex-induced erection does not occur, but an erection is possible with psychogenic stimuli. In addition, injury of the efferent autonomic fibers in the true pelvis, e.g. following retroperitoneal surgery, can also cause erection disorders[212,273].

2.2.5.1.13. Epilepsy

Approximately one-half of all epileptic males suffer from hypogonadism with disorders of libido and sexual potency. Antiepileptic medication appears to be a key factor in the development of hypogonadism. Antiepileptics induce an increase in SHBG and promote the conversion of testosterone to estradiol. This causes a reduction in free testosterone. At the same time, there is a slight increase in gonadotropins, in the context of either the underlying disease or the medication, and therefore the deficiency of free testosterone relative to this reduction in free testosterone is not compensated by increased testicular production. It is difficult to distinguish the effect of the disease from that of the medication, since all investigations are carried out under antiepileptic treatment.

The treatment of hypogonadism with testosterone is not contraindicated in epilepsy. Nevertheless, the dose of antiepileptic might need adjusting, as hepatic metabolism can be affected by androgens. This can be done by monitoring serum drug levels. In addition, it is suspected that estradiol may promote spasmodism, and hence the frequency of convulsions. Since the serum estradiol level may increase on testosterone substitution, testosterone is available in combination with an aromatase inhibitor. In a very small study of five epilepsy patients, simultaneous therapy with testosterone and testolactone, an aromatase inhibitor, significantly reduced the frequency of convulsions while stimulating potency and libido[153].

2.2.5.1.14. Cystic fibrosis and Young's syndrome

The autosomal-recessive inherited disorder of cystic fibrosis is the most common hereditary metabolic disease in Germany. It is caused by a disorder of sodium chloride absorption and water secretion from exocrine glands, with increased viscosity of secretions. Many of the patients affected with this condition now reach adulthood as a result of early intensive therapeutic measures, and therefore some wish to have children. Unlike women with the condition, nearly all men with cystic fibrosis are infertile. They usually have obstructive azoospermia due to hypoplasia or aplasia of the epididymis and vas deferens. The cause of the hypoplasia or aplasia is unclear. It may be caused by degeneration owing to the secretion being highly viscous. The function of the vesicle gland may also be disturbed. Testicular function itself is not impaired, provided that there is no basic restriction of pulmonary function. Therefore, men with cystic fibrosis rarely suffer from testosterone deficiency. However, the onset of puberty is often delayed in boys with the condition, as a result of the severe systemic disease and absorption disorder related to pancreatic insufficiency. The infertility can only be corrected by surgery, possibly by the insertion of artificial spermatocele. However, the chances of fathering children are slight[346].

Young's syndrome, which is rare, is a variant of cystic fibrosis. It can be differentiated from cystic fibrosis by a normal sweat test and unimpaired pancreatic function. As in cystic fibrosis, there is a post-testicular defect; however in Young's syndrome it is due to an obstruction of the epididymis by highly viscous secretions. Young's syndrome is obviously a less pronounced form of cystic fibrosis, and therefore the hypoplasia and aplasia of the epididymis in cystic fibrosis and the obstruction of the epididymis in Young's syndrome present a pathophysiological continuum. Testicular function itself is intact - there is neither testosterone deficiency nor a disturbance of spermatogenesis. Secretolytics were unable to improve the azoospermia in a small group of patients, and therefore the infertility can only be corrected by surgery[145].

2.2.5.2. Medication, radiation, drugs and environmental toxins

A number of medications, physical influences (radiation), drugs and environmental toxins can impair testicular function. Therefore, taking an appropriate case history, including inquiring about workplace conditions and the use of chemicals, will point the way towards a differential diagnosis.

2.2.5.2.1. Medications

Cytostatics

Cell division in the context of spermatogenesis involves a high rate of DNA and RNA synthesis, which is why all medicines that inhibit cell division and nucleic acid synthesis reduce the production of sperm. This applies particularly to cytostatics and antibiotics. Cytostatics can also induce irreversible sterility, depending on the dose. As it is not possible to predict with any certainty the extent and reversibility of fertility damage induced by cytostatics, and this damage cannot be influenced by any other measures at present, it is absolutely essential that the patient be informed about this before treatment starts. Semen should be analyzed to document the pre-therapeutic status. In addition, patients who have not yet completed their families

Substance	Production or motility of sperm	Testosterone production or effect	Libido	Potency	Ejaculation	Gynecomastia	Frequency	Mechanism
Antihypertensives								
α-Methyldopa		↓ (?)	↓	↓	↓	yes	25-50 %	hyperprolactinemia, α-receptor stimulation, sedation
Betablockers				↓			14 %	unknown, possible reduction of penile blood flow
Bethanidine, guanethidine			↓	↓	↓		25-60 %	inhibition of noradrenaline release and reuptake
Clonidine			↓	↓			24 %	central α-receptor stimulation
Dihydralazine				↓ (?)			very rarely	vasodilatation of vessels of the penis
Labetolol				↓	↓		rarely	α- and β-receptor blockade with vasodilatation
Phenoxybenzamine				↓ (?)	↓		frequently	α-receptor blockade with vasodilatation
Phentolamine				↓ (?)	↓		frequently	α-receptor blockade with vasodilatation
Prazosine				↓	↓		rarely	postsynaptic α-receptor blockade with vasodilatation
Reserpine		↓ (?)	↓	↓			1 % (?)	central inhibitor, sedation, depression, hyperprolactinemia
Diuretics								
Amiloride			↓	↓		yes (%)	< 3 %	androgen receptor blockade (?)
Chlortalidone			↓	↓				unknown
Hydrochlorothiazide				↓			16 %	unknown
Spironolactone		↓	↓	↓		yes	30–50%	androgen receptor blockade, inhibition of steroidogenesis
Cardiac drugs								
Digitalis		↓	↓	↓		yes		similar to estrogens
Disopyramide				↓		yes	rarely	anticholinergic effect
Mexiletene			↓	↓		yes (?)	< 1 %	unknown
Verapamil		↓ (?)		↓		yes (?)		reduced LH/FSH secretion
CNS drugs								
Amphetamine			↓					central indirect sympathomimetic
Benzodiazepine			↓	↓			rarely	sedation
Butyrophenone			↓	↓			rarely	hyperprolactinemia, parasympatholytic, sedation
Glutethimide			↓	↓	↓			parasympatholytic effect
Lithium				↓	↓		very rarely	reduced dopaminergic tone
MAO-inhibitors	↓ (?)			↓	↓		frequently	delayed reuptake of noradrenaline
Phenothiazine			↓	↓	↓		to 60 %	α-receptor blockers, parasympatholytic effect, hyperprolactinemia, reduced LH/FSH secretion
Tricyclic antidepressants			↓	↓	↓		25 %	hyperprolactinemia, parasympatholytic effect

Anticholinergics								
Atropine and analogs (also eyedrops!)				↓	↓			parasympatholytic effect
Hormones								
Anabolics/ androgens	↓					yes		suppression of gonadotropins
Estrogens and gestagens	↓	↓	↓	↓		yes		suppression of gonadotropins, inhibition of androgen synthesis
GnRH agonists	↓	↓	↓	↓	↓			reduced LH/FSH secretion
Others								
Allopurinol				↓				unknown
Aminoglute-thimide		↓	↓	↓		yes		inhibition of androgen synthesis
Cimetidine			↓	↓		yes	to 50%	competitive androgen receptor blocker
Clofibrate			↓	↓				androgen receptor blocker (?)
Colchicine	↓						fre-quently	inhibition of cell division
Cyproterone acetate		↓	↓	↓			fre-quently	androgen receptor blockade
Ethionamide				↓		yes	fre-quently	unknown
Etomidate		↓	↓	↓		yes		inhibition of androgen synthesis
Flutamide		↓	↓	↓			fre-quently	androgen receptor blockade
Cortico-steroids		↓	↓	↓				suppression of gonadotropins
Isoniazid				↓		yes	fre-quently	unknown
Ketokonazole		↓	↓	↓		yes	fre-quently	inhibition of androgen synthesis
Levamisole	↓							unknown
Levodopa			↓	↓				parasympatholytic effect
Metoclopra-mide		↓	↓	↓				hyperprolactinemia
Metronidazole			↓					unknown
Nitrofuran-toin					↓			unknown
Phenytoin						yes		increased SHBG production
Salazosulfa-pyridine	↓	↓					fre-quently	unknown, toxic effect (?)

Table 2.15: Type and mechanism of medication effects on testicular function, libido, potency, ejaculation and development of gynecomastia.
CNS = central nervous system; MAO = monoamine oxidase; LH = luteinizing hormone; FSH = follicle stimulating hormone; SHBG = sex hormone-binding globulin; GnRH = gonadotropin releasing hormone.

should attempt to undertake cryoconservation of sperm (☞ Table 2.15).

Unlike the germinal epithelium, the Leydig cells are largely resistant to cytostatics. Androgen deficiency seldom develops after cytostatic therapy. At best, there is compensated Leydig cell insufficiency which is revealed by increased serum LH levels while the serum testosterone remains normal.

Because of the way they impair DNA synthesis, alkylated chemotherapeutics (including busulfan, chlorambucil, cyclophosphamide, melphalan) cause particularly pronounced damage to spermatogenesis with severe oligozoospermia or azoospermia. The effect of these substances on the germinal epithelium is dependent on the total cumulative dose and duration of therapy. Above 400 mg, chlorambucil causes reversible, and above approximately 2600 mg irreversible, azoospermia. If the total dose of cyclophosphamide administered exceeds 18 g, this will always result in azoospermia.

Antimetabolites (e.g. methotrexate, 5-fluorouracil) impair spermatogenesis to a much lesser extent. Only one case of reversible oligozoospermia has been reported even after high doses of methotrexate. The vinca alkaloids (vincristine, vinblastine) cause reversible inhibition of mitotic division of spermatogonia. In addition, they bind to proteins of the microtubules of the sperm flagellum and can therefore affect sperm motility. The impaired spermatogenesis caused by a combination of cisplatin, vinblastine, bleomycin and doxorubicin or adriamycin is also reversible. In the first year following therapy all patients experience azoospermia. However, this is almost always reversible. Between 25 and 32 % of patients have fathered children by the third year after chemotherapy. After ABVD (adriamycin, bleomycin, vinblastine and dacarbazine), 35 % of men present with reversible azoospermia[75,330].

Other medications

Barbiturates as well as verapamil cause a reduction in serum gonadotropin levels. Whereas barbiturates impair GnRH secretion, verapamil reduces the GnRH-induced influx of calcium into the cells of the hypophysis.

Exogenous anabolics and androgens suppress the gonadotropins LH and FSH by negative feedback. This reduces intratesticular testosterone and, together with the FSH deficiency, causes infertility. Libido and sexual potency are not affected. This mechanism has been exploited in clinical studies to develop a male contraceptive. Intake of estrogens, which is usually accidental (cosmetics, vaginal cream), also suppresses gonadotropins and causes gynecomastia and infertility, combined with impaired libido and sexual potency; this is because the androgen effect is absent. Continuous administration of GnRH agonists, which are used in the treatment of carcinoma of the prostate, cause a reduction in hypophyseal GnRH receptors, and LH and FSH are therefore no longer secreted. This leads to infertility and androgen deficiency with loss of libido and sexual potency.

Quite a few medications block the androgen receptor without displaying any intrinsic activity (☞ Table 2.15). The main symptoms are gynecomastia and impaired libido and potency. This group of drugs includes cimetidine, spironolactone, flutamide and cyproterone acetate. In addition, spironolactone, imidazole derivatives (e.g. etomidate, aminoglutethimide) and ketoconazole decrease testosterone production by blocking testosterone synthesis. Moreover, a number of antihypertensives and CNS drugs cause potency and ejaculation disorders[194].

2.2.5.2.2. Exposure to radiation and heat

Ionizing radiation

Unlike the ovary, the germinal epithelium is extremely sensitive to radiation. Fractionated irradiation causes more severe damage than a single dose of the same size. Proliferating spermatogonia are particularly sensitive. As they mature, spermatids become less sensitive to radiation. The above observations explain why changes occur in the semen earlier at high doses than at low doses (☞ Table 2.16). Serum FSH increases as a reflection of the impaired spermatogenesis. Radioiodine therapy of differentiated thyroid carcinoma may also impair spermatogenesis. A total dose of 3700 MBq (100 mCi) appears to be sufficient to do this.

Radiation dose (Gy)	Effect	Recovery
0.1-0.3	oligozoospermia, FSH ↑	complete
0.3-0.5	after 4-12 months azoospermia, FSH ↑	complete
0.5-1.0	after 3-17 months azoospermia, FSH ↑↑	possible
1.0-2.0	after 2-15 months azoospermia, FSH ↑↑	possible
2.0-3.0	after 2 months azoospermia, FSH ↑↑↑	none
> 3.0	azoospermia, FSH ↑↑↑, usually temporary decrease in testosterone or compensated Leydig cell insufficiency	none

Table 2.16: Dose-dependent effects of ionizing radiation on testicular function[194]. FSH = follicle stimulating hormone.

Testicular endocrine function is considerably more resistant to radiation than spermatogenesis. Up to a dose of 0.2 Gy, there is a transient decrease in the serum testosterone level, which is probably due to the reduced circulation of the testes. Hence, the gonadotropins increase. At a dose of 8 Gy and above, there may be irreversible insufficiency of the Leydig cells.

Exposure to heat

In humans, the temperature of the scrotum is 1.5-2 °C below core body temperature. This is possible not only because of its position outside the abdomen, but also because the skin of the scrotum is very thin (no subcutis) and is able to give off large amounts of heat over its entire surface area. Disorders of temperature regulation, as in malpositioning of the testes or varicocele, are associated with infertility disorders of spermatogenesis. It is not known to what extent exposure to heat at work or in leisure time contributes to infertility. However, people who use saunas do not experience an increased rate of fertility disorders[182,205].

2.2.5.2.3. Drugs

Not only does alcohol damage testicular function by damaging the liver, it also has a direct toxic effect on the testes. Acute alcohol consumption in healthy males causes a reversible decrease in serum testosterone after a delay of only a few hours. There is a simultaneous increase in serum LH, indicating transient primary hypogonadism. Alcohol inhibits the activity of an enzyme of testosterone biosynthesis (3β-hydroxysteroid dehydrogenase), thus causing androgen deficiency. There is an initial compensatory rise in LH which returns to normal again in chronic alcohol abuse because of a disturbance in GnRH secretion. Then, serum testosterone is reduced and serum LH normal.

Smoking normal tobacco (e.g. cigarettes) does not appear to have any essential effect on testicular function. The few studies that can be evaluated point to very slightly reduced sperm count and motility in smokers.

Tetrahydrocannabinol (THC), the most important psychotropic substance in marijuana, can cause infertility and androgen deficiency if used regularly. THC inhibits secretion of GnRH reversibly, causing secondary hypogonadism. In addition, semen findings from consumers of marijuana are frequently pathological, showing oligoasthenoteratozoospermia. This indicates a direct (toxic?) effect on spermatogenesis beyond the central effect. In animal experiments, tetrahydrocannabinol has a mutagenic effect and causes chromosome aberrations and aneuploids.

Opiates (morphine, heroin, methadone) interfere with endorphins which promote GnRH secretion, and therefore GnRH secretion falls. Cocaine impairs GnRH secretion, probably also via centrally mediated mechanisms[4,49,364,413].

2.2.5.2.4. Environmental toxins

Reproductive toxicology studies have only been intensively carried out over the past few years. Therefore, although many environmental toxins, heavy metals and chemicals are suspected of impairing fertility, this has only been confirmed for lead, carbon disulfide (CS_2) and the pesticide dibromochloropropane (DBCP) (☞ Table 2.17). As reproductive toxicology studies have only been conducted on a few of the millions of environmental toxins, chemicals and heavy metal compounds, it is still not clear how many and which substances are toxic to gonads. Toxins can affect any part of the hypothalamus-hypophysis-testes axis, and they are also based on several different mechanisms. Therefore, lead causes toxic damage to the germinal epithelium of the testes and a disorder of the hypothalamus-hypophysis interaction. The significance of environmental toxins in idiopathic infertility in man is not clear. However, more recent investigations indicate a connection[104,173,185].

Proven negative effect	carbon disulfide (CS_2), dibromochloropropane (DBCP), lead
Suspected negative effect	acrylamide, arsenic, ethylene dibromide, boron, cadmium, chlordecone (kepon) dichlorodiphenyltrichloroethane (DDT), dimethylaminopropionitrile (DMAPN), dimethyldichlorovinyl-phosphate (DDVP), dimethylmethylphosphonate (DMMP), 1,3-dinitrobenzol, dioxin, ethylene dibromide (EDB), solvents (toluene, benzene, xylene), manganese, polychlorinated biphenyls (PCBs), mercury, carbon tetrachloride, toluene diamine (TDA), trichlorethylphosphate, tris (2,3-dibromopropyl) phosphate (TRIS), vinyl chloride

Table 2.17: Environmental and industrial substances proven or suspected to have a negative effect on testicular function. The suspicions are largely based on the results of animal studies. Whether this effect transfers to humans has not been proven.

Diagnostic work-up of hypogonadism

3. Diagnostic work-up of hypogonadism

Patients with hypogonadism usually seek medical advice when they are unable to father children, or suffer from erectile dysfunction, loss of libido or incomplete puberty. A detailed case history and physical examination will help towards a diagnosis. The main symptoms and suspected diagnosis can then be confirmed by laboratory tests. In the light of these results, more extensive investigations such as osteodensitometry, MRI, chromosome analysis or genetic testing may be necessary.

3.1. Case history and physical examination

The case history offers the opportunity to ask questions about time of onset and extent of symptoms and accompanying or preceding circumstances (e.g. viral infections or general illnesses). Headaches and visual disorders indicate a hypophyseal process, and anosmia indicates Kallmann's syndrome. Pathological fractures or increasing bone pain might indicate osteoporosis. Previous conditions (e.g. delayed puberty), surgery (e.g. orchidopexy) and therapeutic measures (e.g. hormone therapies) are significant, as is a family history (relatives with hypogonadism, parents with delayed puberty) and a medication history. A full and frank sexual history should be obtained. In fact, many patients are grateful when this aspect is addressed and they are able to express their feelings on this subject. They should be asked about frequency of sexual intercourse, libido, spontaneous morning erections and satisfaction with their sex life.

The physical examination comprises a general examination including measurement of blood pressure. Depending on age, the assessment will take into account body proportions (e.g. excess eunuchoid height, female-type fat distribution), virilization (secondary hair growth, facial hair, musculature) and the presence of gynecomastia. It is important to examine pubic hair to evaluate the stage of puberty and whether the normal male diamond pattern is present on the abdomen. An adult penis should be 6 cm long and approximately 3 cm in girth. The correct position of the urethral opening should be assessed after drawing back the foreskin. Hypospadia and epispadia indicate a disorder of fertility or sexual differentiation. The folds and pigment of the scrotum should be examined to assess whether they are normal. The testes should be carefully palpated to evaluate size and consistency. Lumps or pain could indicate a malignoma and should be evaluated by ultrasound. Normal testes of men of Caucasian origin are 4 cm or more in length and 12 ml or more in volume (Prader's orchidometer, Schirren's circle and testicular sonography). Fluctuating masses indicate hydrocele, which can be verified by illuminating with a torch in a dark room or carrying out sonography. Varicocele and varicose enlargements of the spermatic vein tend to occur on the left side. They are particularly noticeable during the Valsalva maneuver. Hardening of the epididymis or the spermatic cord may correspond to spermatocele or previous inflammation, and should also be verified by sonography. Finally, the prostate should be palpated to evaluate size, consistency and any focal changes[80,422].

General examination:

- Body proportions, e.g. excess eunuchoid growth, accentuated fat distribution as in females
- Virilization (secondary hair growth, facial hair, musculature)
- Gynecomastia
- Pubic hair (distribution, density)
- Penis (size)
- Position of urethral opening (hypospadia/epispadia)
- Scrotum (pigmentation, hydrocele)
- Testes (volume, consistency, lumps)
- Spermatic cord (varicocele)
- Prostate (size, consistency, lumps)

3.2. Hormone analysis

3.2.1. Testosterone, LH and FSH

> The evaluation of male hypogonadism always includes measurement of serum testosterone, LH and FSH. Testosterone reveals information about the endocrine activity of the testes, and LH and FSH serve to evaluate hypophyseal function. Both together allow an etiological classification of hypogonadism.

Increased levels of gonadotropins indicate primary hypogonadism; low levels of gonadotropins combined with low levels of testosterone are indicative of secondary hypogonadism. In addition, increased levels of FSH are considered to indicate inadequate functioning of the germinal epithelium. Apart from the rare exception of androgen resistance, increased LH with normal testosterone indicates Leydig cell insufficiency.

When evaluating serum testosterone, the circadian rhythm of testosterone production must be borne in mind. Owing to increased nightly production of LH, testosterone production reaches its peak in the early morning hours, and then falls again slowly until it reaches a minimum in the evening. Trough levels are 25-40 % below peak levels.

Therefore, blood samples should be taken early in the morning.

Since the daily rhythm becomes less significant with increasing age, when peak levels approach trough levels, the timing of blood samples is not as important in older men as in young men. Reference ranges are listed in Table 3.1.

In androgen deficiency of the older man, which is specifically characterized by a marked decrease in free testosterone, the question is whether total testosterone or free testosterone should be measured. Normally, the routine method for measuring testosterone is to measure total testosterone, i.e. free and protein-bound testosterone. Determination of free or 'dialyzable' testosterone measures the proportion of non-protein-bound testosterone. One very costly method uses dialysis tanks, but commercial kits for measuring free testosterone are also available. However, these are inaccurate at very low concentrations of testosterone and high levels of SHBG[39], and therefore often do not provide reliable results. Therefore, the general consensus of opinion is that, until better methods for determining free testosterone are introduced, the sa-

Parameter	Typical reference range	Conversion factors	Our reference range
Testosterone	12-30 nmol/l	nmol/l x 0.2884 = ng/ml	
	350-850 ng/dl	ng/dl x 0.03467 = nmol/l	
	3.5-8.5 ng/ml	ng/ml x 3.467 = nmol/l	
Free testosterone	174-900 pmol/l	pmol/l x 0.3467 = pg/ml	
	60-312 pg/ml	pg/ml x 2.884 = pmol/l	
DHT	1-3.5 nmol/l	nmol/l x 0.2899 = µg/l	
	2.9-10 ng/dl	ng/dl x 0.0344 = nmol/l	
	0.29-1 µg/l	µg/l x 3.44 = nmol/l	
17β-estradiol	25-85 pmol/l	pmol/l x 0.2724 = pg/ml	
	7-23 pg/ml	pg/ml x 3.671 = pmol/l	
SHBG	30-70 nmol/l		
LH	2-10 IU/l		
FSH	1-7 IU/l		
Prolactin	< 20 ng/ml	ng/ml x 44.4 = pmol/l	

Table 3.1: Normal reference ranges for andrologically significant hormones and conversion factors for laboratory parameters. The reference ranges are method-specific and should be obtained from the laboratory carrying out the tests. A column has been left in the table for these.
DHT = 5α-dihydrotestosterone; SHBG = sex hormone-binding globulin; LH = luteinizing hormone; FSH = follicle stimulating hormone.

fer, more accurate measurement of total testosterone should be used[39]. However, to determine free testosterone, an index of free testosterone can be calculated from the values for total testosterone, SHBG and total protein (or albumin). This tallies largely with the actual value for free testosterone (determined by equilibrium dialysis)[278,427]. The International Society for the Study of the Aging Male (ISSAM) has a calculator on its internet page (http://www.issam.ch/freetesto.htm). Choice of method is less important than a reference range, which should be specifically determined for the method used and also include all age groups, especially older and elderly men. Moreover, the methods should be critically analyzed, since commercially available testosterone kits differ considerably in quality[183].

Food intake does not affect testosterone levels; however, increases in lipid fractions may affect the accuracy of the analysis, and therefore it is preferable to take samples after fasting. Short, intense physical exertion can cause an increase, and exhausting, longer-term physical exercise a decrease in serum levels. Serious disease (e.g. renal failure, myocardial infarction, liver cirrhosis), stress, surgery, excessive alcohol consumption and numerous medications (including corticosteroids) can lead to a decrease in testosterone. Unlike narcotics, nicotine does not affect testosterone.

The hCG stimulation test with human chorionic gonadotropin, which carries biological activity as LH, is used to differentiate between cryptorchidism and anorchidism. The test is standardized to intramuscular administration of 5000 IU hCG. Serum testosterone is determined before (baseline) and 72 h after injection. A 1.5-2-fold increase in serum testosterone confirms the presence of testicular tissue, and in cryptorchidism (undescended testicles) the testes should then be sought using imaging techniques, e.g. MRI. If there is no increase, then no testicular tissue is present.

Gonadotropins are secreted in a pulsatile fashion, and therefore show short but sharp fluctuations. These are less pronounced with FSH than with LH, owing to its longer half-life. To prevent misinterpretation because of drawing a sample during a peak or pronounced trough in secretion, three samples can be taken over a period of 60-90 min at 20-30 min intervals. These samples are then combined in a test-tube so that a representative value can be obtained from one analysis.

Increased LH values indicate primary (testicular) hypogonadism, and low or low-normal values combined with decreased testosterone secondary (hypophyseal-hypothalamic) hypogonadism (☞ Table 3.3). In contrast, FSH is subject to negative feedback not only from the sex steroids, but also from inhibin, which is produced by the Sertoli cells. Hence, increased FSH reflects infertility and disorders of spermatogenesis. Determination of FSH has largely superseded testicular biopsy. If high FSH levels are combined with small testes (less than 6 ml in volume) and azoospermia, Klinefelter's syndrome should always be suspected. If the testes are larger and there is azoospermia or oligozoospermia, a high FSH indicates a primary disorder of spermatogenesis. In the case of azoospermia with normal-sized testes and normal FSH one should suspect closure of the efferent seminal ducts.

• Neoplasias - testicular tumors - (Leydig cell and Sertoli cell tumors) - chorionic carcinomas - hepatomas • Primary hypogonadism • Liver cirrhosis • Hyperthyroidism • Adrenal adenomas and carcinomas • Obesity • Androgen-resistance syndrome • Estrogen receptor defect • Medication - estrogens - antiandrogens - antiestrogens - hCG

Table 3.2: Causes of increased serum concentrations of 17β-estradiol. hCG = human chorionic gonadotropin.

The GnRH test is used to distinguish between low-normal and reduced gonadotropin levels. It is also indicated to distinguish between hypophyseal and hypothalamic hypogonadism and to differentiate between constitutional delayed puberty and idiopathic hypogonadotropic hypogonadism (IHH). To perform the test, 100 µg of gonadotropin releasing hormone (GnRH, luteinizing hormone releasing hormone (LHRH)) is injected intravenously. Blood samples are taken before (baseline) and 45 min after injection to determine gonadotropin levels. In healthy men LH increases 1.5-2-fold, whereas FSH is only slightly stimulated. An inadequate increase even after pretreatment with GnRH indicates hypophyseal insufficiency. However, if the values do increase significantly from their low baseline levels in a patient with testosterone deficiency, the hypogonadism is hypophyseal in origin.

3.2.2. Additional hormone parameters

Other parameters will aid diagnosis. In secondary hypogonadism, prolactin levels are always measured to exclude/detect prolactinoma. When evaluating serum prolactin, the influence of a number of medications (e.g. CNS drugs, metoclopramide) as well as stress should be taken into account. Therefore, if prolactin levels are only slightly increased (< 2 times the reference range), repeated samples should be taken to exclude stress-induced hyperprolactinemia (☞ Table 3.1).

If there is clinical evidence of a mass in the region of the hypophysis, further endocrinological diagnosis will be required regarding thyrotropic, adrenocorticotropic and somatotropic hypophyseal function, since in addition to hypogonadism there may also be hypothyroidism or adrenal insufficiency. Ideally, the patient should be referred to an endocrinologist for specialist diagnosis of hypophyseal function.

If there is gynecomastia, in addition to prolactin, serum 17β-estradiol levels should be measured. Unusually severe osteoporosis in a young man will additionally necessitate determination of estradiol levels to detect aromatase deficiency or estrogen receptor defect, both of which are very rare. Estradiol will not normally need to be measured in other patients with hypogonadism as this is not necessary for diagnosis or treatment. Estradiol levels are also not required in andropenia of the older man as estrogen treatment is not given, and the decision whether or not to use testosterone therapy will not depend on estradiol levels. Causes of increased estrogen levels are listed in Table 3.2.

T ↓, LH ↑, FSH ↑	suspicion of primary hypogonadism (e.g. Klinefelter's syndrome, anorchidism)
T ↓, LH ↓, FSH↓	suspicion of secondary hypogonadism, GnRH test, if necessary hypophyseal diagnosis (e.g. IHH, Kallmann's syndrome, prolactinoma, hypopituitarism)
T n, LH ↑, FSH n or↑	compensated Leydig cell insufficiency, e.g. Klinefelter's syndrome, radiotherapy, chemotherapy, viral orchitis, undescended testicles)
T n, LH n, FSH ↑	suspicion of disorders of spermatogenesis (e.g. Sertoli-cell-only syndrome, radiotherapy, chemotherapy, viral orchitis, undescended testicles)
T n, LH n, FSH n	no testicular endocrine insufficiency, disorders of spermatogenesis with infertility not excluded (e.g. idiopathic infertility, globozoospermia, nonmotile cilia syndrome), in severe oligozoospermia and particularly obstruction of the efferent seminal ducts
T ↑, LH ↑, FSH ↑	suspicion of androgen resistance; contact specialist for gene diagnosis

Table 3.3: Typical results from hormone analysis in male hypogonadism (see Table 2.2). LH = luteinizing hormone; FSH = follicle stimulating hormone; T = testosterone; ↓ decreased; ↑ increased, n = normal; IHH = idiopathic hypogonadotropic hypogonadism.

Determination of SHBG and free testosterone is unnecessary in the majority of patients with hypogonadism. However, SHBG may need to be determined in patients with disorders of SHBG production or other factors (see Chapter 1.3.2). Specific radioimmunoassays are used today to determine SHBG.

Serum levels of 5α-dihydrotestosterone (DHT) remain at a relatively constant 10 % of the serum testosterone level in a man with normal 5α-reductase function. Therefore, it is sensible to determine DHT only if one suspects male pseudohermaphroditism resulting from a congenital deficiency of 5α-reductase. However, this requires complete separation of DHT and testosterone, which is difficult because the molecules are so similar and results are therefore often inaccurate.

There are also no clinical consequences from determining inhibin, which is produced by the Sertoli cells and which has a negative feedback effect on FSH. Therefore, there is no point in determining inhibin, except for scientific purposes.

The adipocyte hormone leptin is increased in androgen deficiency but returns to normal with testosterone substitution. Therefore, leptin is a suitable parameter for evaluating testosterone substitution, but it is not routinely used because it is influenced by other factors (e.g. food intake, fat mass, insulin).

3.3. Other biochemical parameters

Androgens affect fat metabolism, erythropoiesis and liver metabolism, so these parameters are checked before the start of testosterone therapy and checked regularly during therapy.

If a patient has been suffering from androgen deficiency for some time, he may have androgen-deficiency anemia, which can be recognized by a low hematocrit, reduced red blood cell count and reduced concentration of hemoglobin. These parameters usually return to normal during the course of testosterone substitution. Therefore, serum hemoglobin and hematocrit are important parameters for evaluating the quality of testosterone substitution in hypogonadism.

The administration of testosterone may affect lipid metabolism. Therefore, before and during therapy it is recommended that total cholesterol, LDL cholesterol and HDL cholesterol are measured. Men with a proven indication for testosterone substitution, and who have pathological HDL cholesterol or LDL cholesterol values, should be monitored to ensure that they are receiving the correct dose. Administration of a cholesterol synthesis inhibitor may need to be considered.

Anabolic steroid abuse and testosterone overdose, e.g. if doses are too high or administration intervals too short, can be detected by an increased hematocrit (> 50 %), increased hemoglobin concentration (> 18 g/dl) and low HDL cholesterol (< 35 mg/dl). As many anabolics are not detected when measuring testosterone, these parameters are an important indicator if a person is concealing the fact that he is taking them.

Unlike other androgens and anabolics, testosterone does not have an adverse effect on liver function when administered in normal doses. Nevertheless, bilirubin and transaminases (glutamicoxaloacetic transaminase (GOT), glutamicpyruvic transaminase (GPT)) should be measured before the start of therapy to detect any liver disorder that may have gone undetected. Bilirubin may be increased in predisposed men (e.g. those with Gilbert-Meulengracht syndrome), but this does not usually mean that therapy has to be discontinued. Follow-up checks are carried out 3 months after the start of treatment and then annually.

The effect of testosterone treatment on the prostate becomes more important with age, which is why prostate-specific antigen (PSA) is measured from the age of about 45-50 years. During treatment there is often an increase in PSA, but this should not exceed the upper limit of normal. If PSA increases above this level, treatment should be stopped until the finding has been investigated (sonography or biopsy). Nowadays age specific normal ranges are provided by the laboratory, which depend on the kind of assay used.

Determination of other biochemical parameters is dependent on the case history and findings from the general physical examination. In diabetes mellitus, control should be monitored by the proportion of glycosylated hemoglobin (HbA_{1c}), and in renal insufficiency retention parameters should be measured (creatinine, urea) and in liver disease, in addition to transaminases, parameters of synthesis (e.g. cholinesterase, Quick's test, albumin).

3.4. Semen analysis

Semen analysis forms an important part of andrological diagnosis. In men who have so far been unsuccessful in having children it is an essential component of the investigation into testicular function. However, semen analysis also provides important information in suspected endocrine disorders of testicular function, e.g. Klinefelter's syndrome, other forms of primary hypogonadism and mild androgen resistance. Semen analysis is not normally performed when investigating delayed puberty or androgen deficiency in the elderly.

Nowadays, semen analysis should be carried out exclusively according to the standardized and generally accepted method laid down by the World Health Organization (WHO). The WHO laboratory handbook describes all the test methods and laboratory analyses in detail, it is not described here[433]. However, the classification and normal values are given for information in Tables 3.4 and 3.5.

Classification	Description
Normozoospermia	normal finding
Oligozoospermia	< 20 x 10^6 spermatozoa/ml
Asthenozoospermia	< 50% spermatozoa with progressive motility
Teratozoospermia	< 30% spermatozoa with normal morphology
Oligoasthenoterato-zoospermia (OAT)	combined disorder of all three parameters
Azoospermia	no spermatozoa in semen
Aspermia	no ejaculate

Table 3.4: Classification for semen findings[433].

Parameter	Value
Volume of ejaculate	> 2 ml
pH	7.2-8.0
Concentration of sperm	> 20 x 10^6 spermatozoa/ml
Sperm count/ejaculate	> 40 x 10^6 spermatozoa
Motility	> 50% spermatozoa with progressive motility (categories a and b) or > 25% with fast progression (Category a)
Morphology	> 30% spermatozoa with normal morphology
Vitality	> 75% vital sperm
Leukocytes	< 1 x 10^6/ml
α-Glucosidase	> 20 mU/ejaculate
Fructose	> 13 μmol/ejaculate
Zinc	> 2.4 μmol/ejaculate

Table 3.5: Reference ranges for semen during a preparatory period of 48 hours to 7 days[433].

3.5. Testicular biopsy

Nowadays, biopsy of the testes is performed only if a tumor is suspected or if efferent seminal ducts are suspected to be closed (azoo- or severe oligozoospermia with normal FSH, normal testicular volume and low epididymis marker). If FSH is increased and semen findings are pathological, biopsy is superfluous as there are no consequences for the therapy. It is important to consult a pathologist who is experienced in evaluating testicular tissue.

3.6. Sonography of scrotum and prostate

Sonography supplements palpation of the scrotum and enables testis volume to be measured accurately (☞ Figure 3.1). In addition, the internal structure of the testes, focal changes (tumors) and intrascrotal masses (hydrocele, spermatocele) can be identified. A varicocele will show up as an enlargement of the pampiniform plexus (if necessary, after cough provocation or Valsalva maneuver). Previous inflammation of the epididymis will show up as echo-rich changes or calcifications.

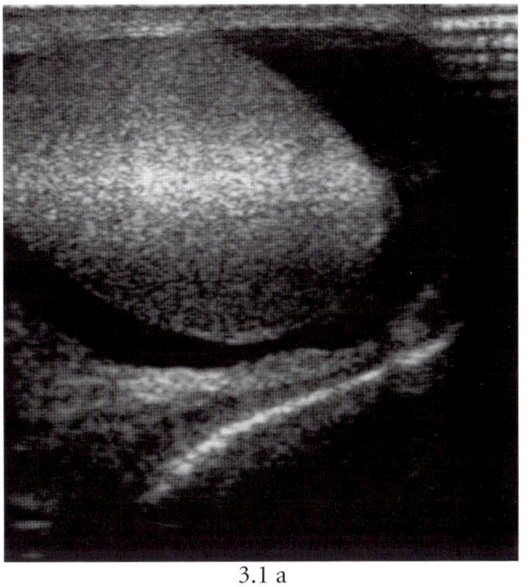

3.1 a

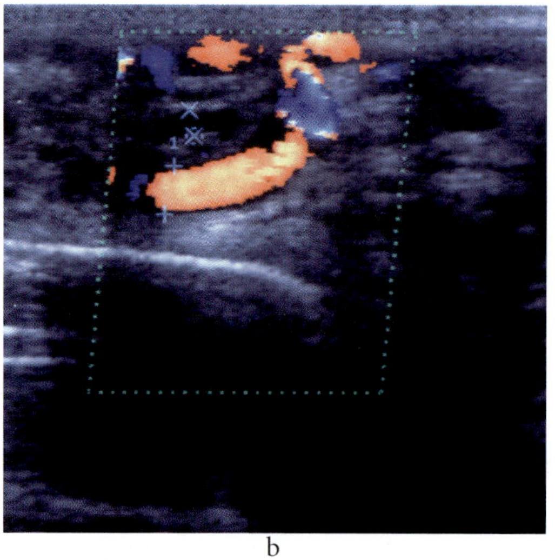

b

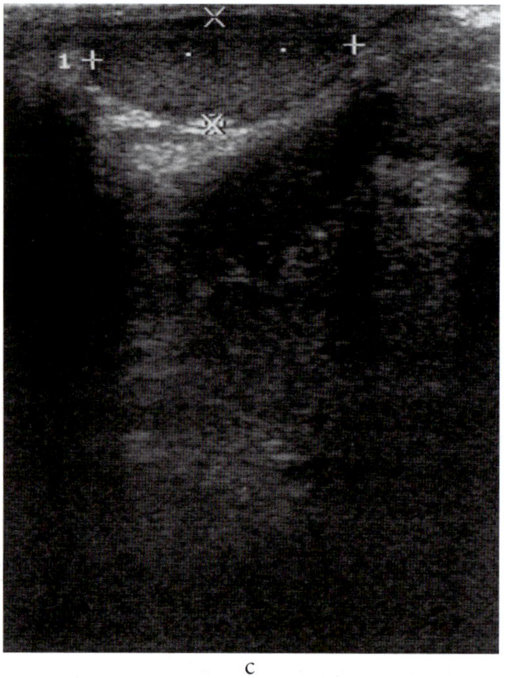

c

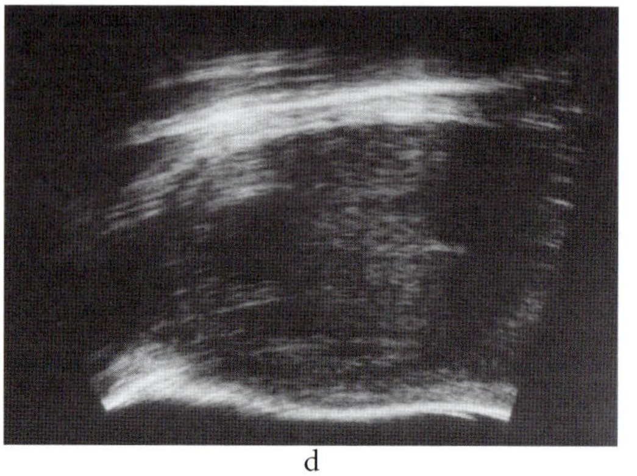

d

Figure 3.1: Examples of testicular imaging (sonography).
(a) Normal testes with pronounced hydrocele.
(b) Varicocele in color duplex sonography showing very convoluted vessels with enlarged diameter.
(c) Undescended testicles in a 14-year-old boy with delayed puberty as a result of idiopathic hypogonadotropic hypogonadism.
(d) Multifocal teratoma of the testis.

Abdominal sonography of the prostate is not as useful as rectal ultrasound (transrectal ultrasound, TRUS) in determining focal changes and exact size. This is particularly true of older men with andropenia where focal changes are especially relevant. TRUS is always carried out on men over 45-50 years of age before androgen therapy starts. Within a few months, testosterone therapy will normalize prostate volume if the prostate was small as a result of androgen deficiency. If androgen therapy is to continue, the TRUS should ideally be repeated annually in order to recognize early signs of excessive growth, benign hyperplasia of the prostate or focal changes with potential malignancy. Any lumps in the prostate must be investigated before androgen therapy is started, if necessary using a punch biopsy (☞ Figure 3.2). If any new changes arise during androgen therapy, treatment should be halted immediately until the matter has been investigated. In benign hyperplasia of the prostate, determination of urinary flow can provide additional, clinically relevant information.

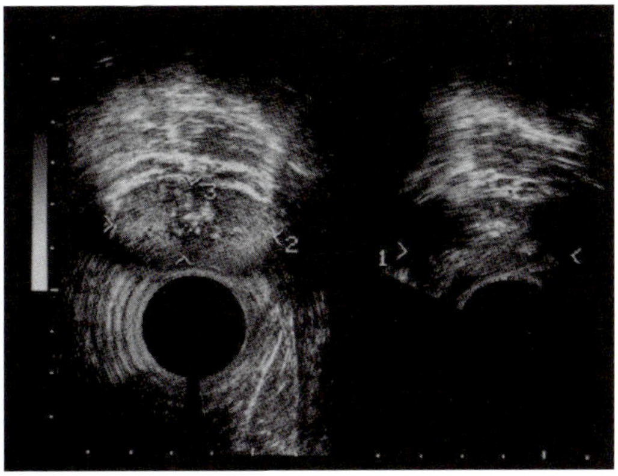

a

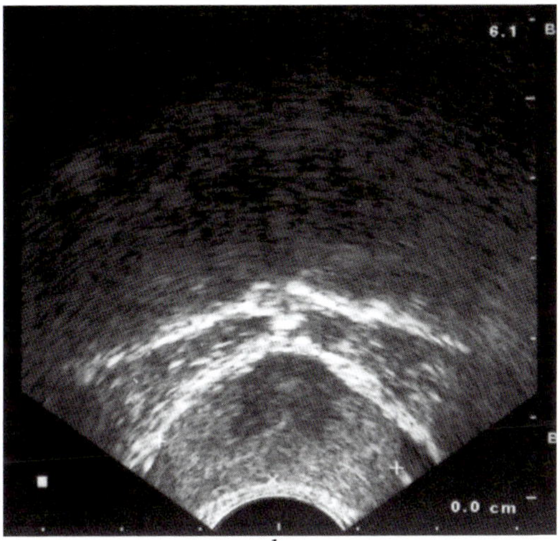

b

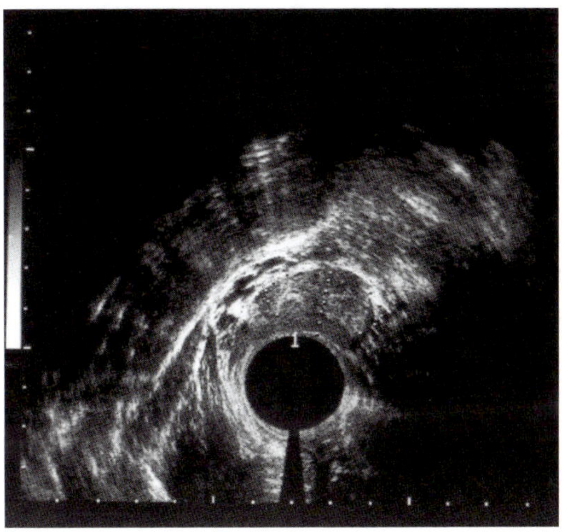

c

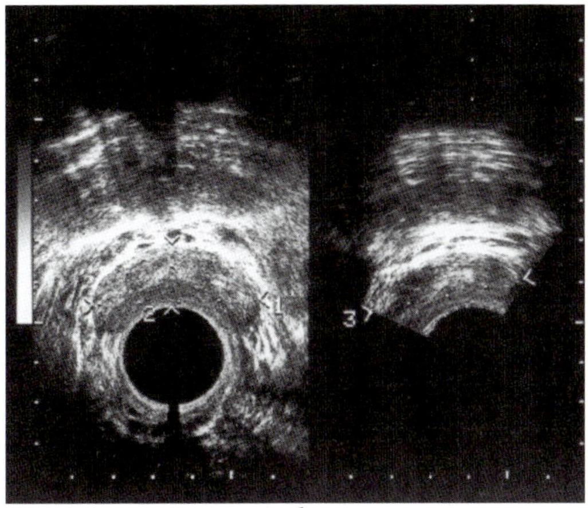

d

Figure 3.2: Transrectal ultrasound of the prostate.
(a) Central calcification indicates previous prostatitis.
(b) Echo-poor mass as incidental finding in a 53-year-old man with Klinefelter's syndrome before testosterone substitution. Punch biopsy shows no evidence of malignancy and situation remains unchanged during testosterone therapy.
(c) Small prostate in a 19-year-old with congenital anorchidism, prostate volume 7.7 ml.
(d) Same patient as Figure 3.2c after 7 months of testosterone substitution treatment, prostate volume 14 ml.

3.7. Osteodensitometry

All men with hypogonadism should have their bone mineral content/bone density measured. This is useful for evaluating current fracture risk as well as for monitoring the course of testosterone substitution therapy. If osteodensitometry reveals low bone density during testosterone therapy, for example in men with andropenia, this is a sign that treatment needs to be supplemented.

The quality of testosterone substitution can be measured during treatment by changes in bone density. The following are generally accepted and well-validated osteodensitometry methods: quantitative computed tomography (QCT), dual photon absorptiometry (dual-energy X-ray absorptiometry, DEXA) and peripheral quantitative CT (pQCT), although photon absorptiometry offers the added advantage of being able to document body composition (proportion of fat and muscle mass). Sonographic methods for evaluating fracture risk have not yet been sufficiently validated, and are therefore less suitable than established methods in reaching a diagnosis and monitoring therapy. Bone density checks are carried out every 1-2 years (Figure 3.3).

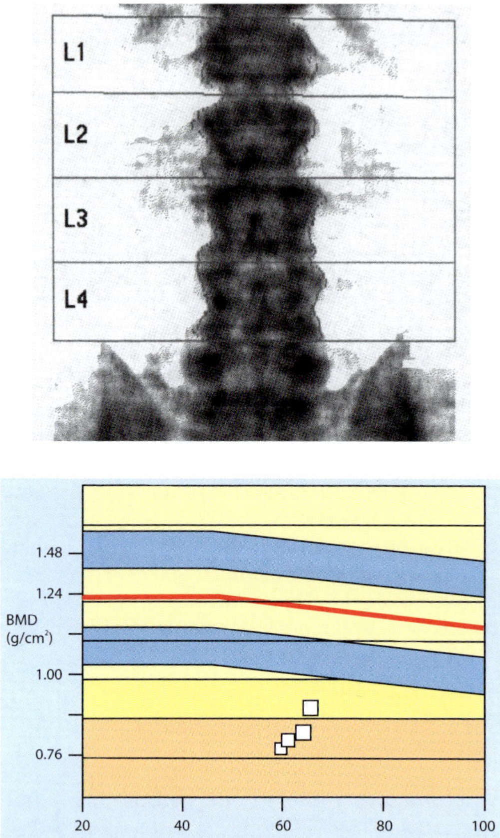

Figure 3.3: Osteodensitometry (dual photon absorptiometry, DEXA) of the lumbar vertebral column in a 60-year-old man with hypogonadism and severe osteoporosis. Bone density, which is significantly reduced before testosterone substitution, increases during therapy until it exceeds the lower normal limit. The mean value is shown by a red line and the area inside the blue bars corresponds to 2 standard deviations.

3.8. Nuclear magnetic resonance imaging of the hypophysis

MRI should be carried out in secondary hypogonadism to detect hypophyseal adenoma and to exclude masses in the hypothalamic-hypophyseal region. MRI is more sensitive than computed tomography (CT). So that even small adenomas, e.g. microprolactinomas, do not elude detection, tomography should be carried out in 2-mm sections with and without contrast medium (gadolinium). Lateral X-ray of the skull and sella target X-ray are now obsolete (☞ Figure 2.7). If there is clinical evidence of impairment of the optic chiasma (impaired vision, bitemporal vision) or macroadenoma, an ophthalmological examination should always be carried out, including perimetry (measurement of visual field).

3.9. Chromosome analysis and biomolecular investigations

If genetic disorders of testicular function or aberrations of chromosome number are suspected, this can be confirmed by appropriate tests in specialist clinics. The combination of very small testes (< 4 ml) with increased FSH and azoospermia generally leads to a diagnosis of Klinefelter's syndrome and cytogenetic testing. This condition can be rapidly confirmed by the detection of Barr bodies (extra X-chromosome) in a swab of mouth mucosa (☞ Figure 3.4). However, a negative finding does not necessarily exclude Klinefelter's syndrome.

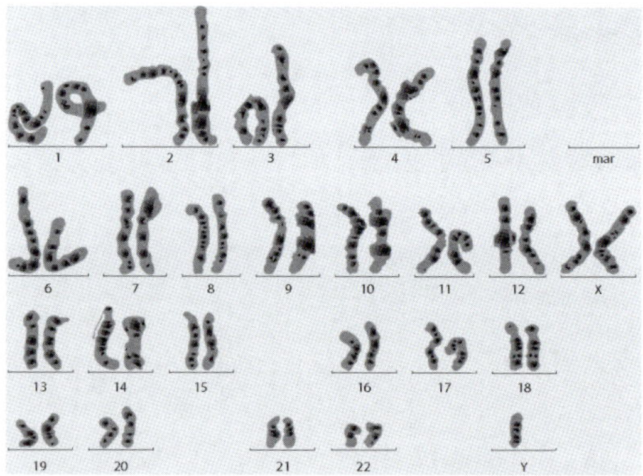

Figure 3.4: Karyogram of a patient with Klinefelter's syndrome. There are two X-chromosomes and one Y-chromosome, so the karyotype is 47,XXY.

Androgen-resistance syndrome, suspected estrogen receptor defect, 5α-reductase deficiency and aromatase deficiency are confirmed by biomolecular determination of gene mutation.

All very rare diseases and syndromes should be referred to specialist endocrinology clinics who can make a rapid diagnosis, which is beneficial to patients.

Androgen treatment options

4. Androgen treatment options

Testosterone is the most important androgen produced by the testes. Approximately 5-7 mg testosterone is synthesized by the healthy testes each day. Therefore, except for a very few rare indications, androgen therapy means administration of testosterone. Treatment of hypogonadism in men in particular is based on the administration of testosterone (☞ Figure 4.1). The aim of therapy is to establish a physiological concentration of serum testosterone which is appropriate for a patient's age in order to correct the androgen deficiency, relieve its symptoms and prevent long-term sequelae.

> Since there are enormous differences in levels of testosterone and the requirement for testosterone at various stages in a person's life, androgen therapy should be adjusted to the age and current reproductive requirements of the patient.

There are now a number of preparations and formulations available which mean that the majority of patients can select a clinically effective product which is appropriate for them.

4.1. Preparations

The liver shows a high capacity for breaking down circulating testosterone. The half-life of free testosterone in the serum is approximately 10 min. Therefore, orally administered testosterone, which is absorbed well from the duodenum, causes only a very short, clinically ineffective increase in serum testosterone. Doses only in excess of 200 mg testosterone will exceed the metabolic capacity of the liver and reach the target organs via the systemic circulation[197].

> This is why in androgen therapy non-oral formulations are chosen, or chemically modified androgens are used, which are metabolized less quickly than native testosterone (☞ Figure 4.2). Therapeutic androgens can be classified according to route of administration or chemical structure (☞ Table 4.1).

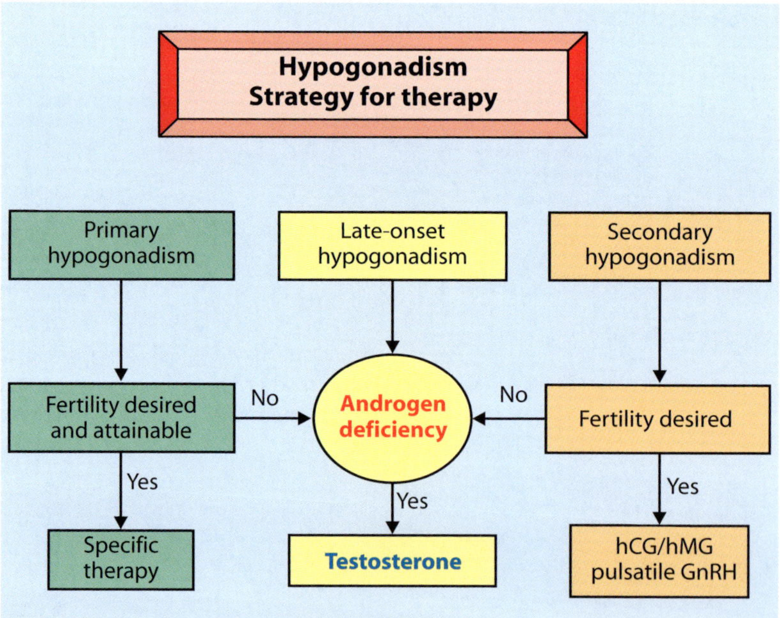

Figure 4.1: Strategy for the treatment of hypogonadism in man. Treatment of all forms of male hypogonadism is based on the administration of testosterone. Treatment is only interrupted in favor of other procedures during short periods when fertility is desired.

Classification according to	
Route of administration	
Oral	• mesterolone • testosterone undecanoate • metenolone acetate • methyltestosterone • fluoxymesterone
Intramuscular	• testosterone propionate • testosterone enanthate • testosterone cypionate • testosterone undecanoate • testosterone buciclate • 19-nortestosterone decanoate
Subcutaneous	• testosterone pellets
Transdermal	• testosterone gel • testosterone patches
Chemical structure	
Native testosterone	• testosterone gel • testosterone patches • testosterone pellets
17α-Hydroxyl-ester	• testosterone propionate • testosterone enanthate • testosterone cypionate • testosterone undecanoate • testosterone buciclate • 19-nortestosterone decanoate
1-Alkylated androgens	• mesterolone • metenolone acetate
17α-Alkylated androgens	• methyltestosterone • fluoxymesterone

Table 4.1: Classification of therapeutic androgens.

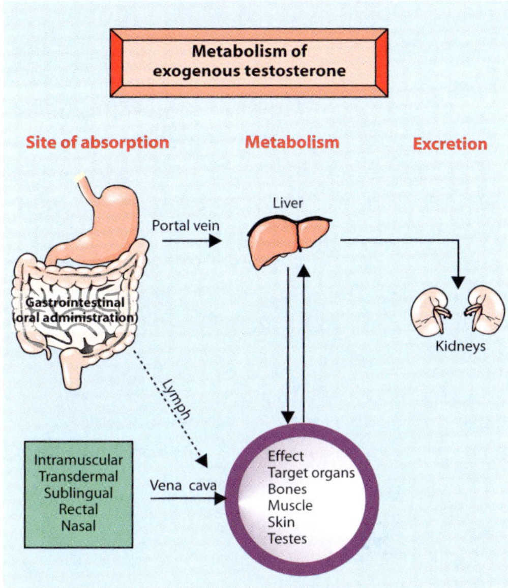

Figure 4.2: Administration of exogenous androgens. Orally administered testosterone is absorbed by the intestines and reaches the liver via the portal vein where it is quickly and almost totally inactivated. Oral testosterone undecanoate is predominantly absorbed by the lymphatic system in the intestines and therefore bypasses hepatic elimination, as in the parenteral route. Therefore it reaches the target organs before the liver. After inactivation and glucuronidation, elimination is via the kidneys.

All conceivable modes of application of testosterone have been evaluated except via the auditory canal; however, for practical reasons, nasal, rectal and conjunctival administration have not become popular[1,18,93]. Transdermal, intramuscular and oral androgen preparations are most widely used (☞ Table 4.2). In some countries, the range of substances additionally includes subcutaneous, sublingual and buccal formulations.

In addition to the development of various formulations, there have been numerous modifications of the chemical structure of the testosterone molecule to extend its clinical efficacy (☞ Figure 4.3).

Thus, alkylation in the 17α-position prevents rapid breakdown in the liver. However, this causes hepatotoxicity, and therefore these preparations are no longer recommended to be used. Esterification of the 17β-hydroxyl group, which also prevents premature metabolism, does not have this side-effect. Esterification with carboxyl groups in the 17β-position increases lipophilia and enables intramuscular administration in an oily solution. This is released slowly from the depot injection - the longer is the aliphatic side-chain, the slower is the rate of delivery. However, esters which enter the circulation display the same short half-life as that of testosterone itself. The prolonged effect is therefore exclusively based on delayed release from the depot injection. Alkylation at position 1 also protects from the 'first-pass' effect of the liver. Removal of the C19 group (19-nortestosterone) alters the metabolism and therefore increases the duration of effect.

Oral	testosterone undecanoate	40 mg capsules
	testosterone	30 mg buccal tablets
	clostebol-acetate	15 mg tablets
Intra-muscular	testosterone enanthate	250 mg ampoules
	testosterone cypionate	100/200 mg ampoules
	testosterone undecanoate	1000 mg ampoules
	testosterone propionate	50 mg ampoules
	clostebol-acetate	10 mg ampoules
	nandrolone decanoate	25/50 mg ampoules
Trans-dermal	testosterone	5 g gel
		2.5 g gel
		12.2 mg body patch
		15 mg scrotal patch

Table 4.2: Commercially available androgens.

Attempts have also been made to separate the androgenic, virilizing effect from the anabolic effect by changing the structure of the testosterone molecule. The protein-anabolic effect could then be exploited in women and children. In addition, attempts have been made to use it in the treatment of osteoporosis, aplastic anemia and cachexia, without the unwanted androgenic effect. A number of so-called anabolic steroids (anabolics) have been synthesized which are supposed preferentially to stimulate erythropoiesis and the build-up of muscle and bone. The anabolic-androgen index (calculated as the ratio of the weight increase of a muscle to the weight increase of the seminal vesicle of the rat being given steroid hormone relative to a reference steroid, usually testosterone) is greater for anabolics than for testosterone, which indicates that anabolics have a stronger protein-anabolic effect. It is now known that all androgens exert their effect via the same androgen receptor, which is why attempts to separate completely the androgen effect from the anabolic effect have not been successful; all anabolic steroids show a varying degree of androgen effect.

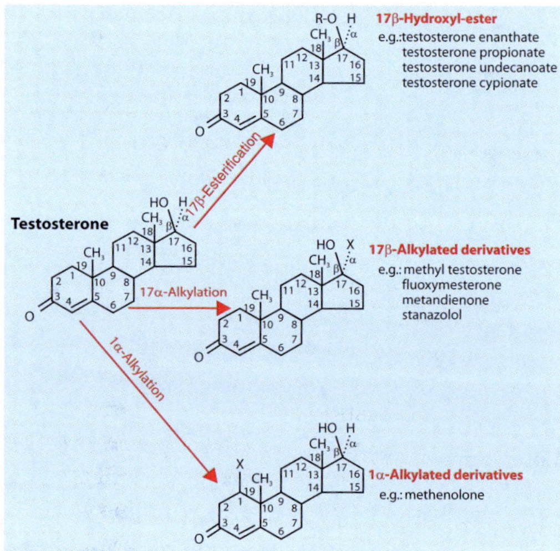

Figure 4.3: Common structural modifications of the testosterone molecule to extend clinical efficacy.

4.1.1. Oral androgens

4.1.1.1. Mesterolone

Mesterolone is a derivative of dihydrotestosterone (1α-methyl-5α-dihydrotestosterone) and occupies a special place among androgen preparations (☞ Figure 4.4).

After oral administration, mesterolone is absorbed well and protected from metabolism in the liver by the methyl group. A single intake of 30 mg produces a mesterolone serum level which is approximately equivalent to the physiological concentration of testosterone. Maximum blood levels are attained after approximately 3 h and the half-life is approximately 7 h. Therefore, to maintain a continuous concentration it is necessary to administer approximately 150 mg per day in 3-4 divided doses.

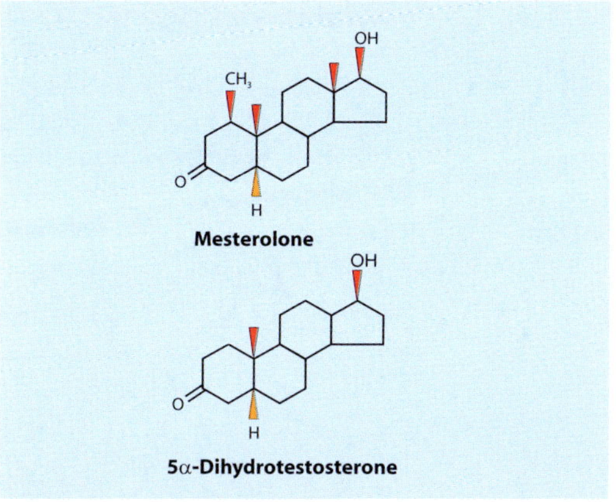

Figure 4.4: Structural formula of mesterolone, a derivative of 5α-dihydrotestosterone.

Mesterolone cannot be converted to estradiol or testosterone by the body. Therefore, it does not show the complete spectrum of effects of testosterone, and is not suitable for substitution in male hypogonadism. Nevertheless, mesterolone is useful in some situations. Hence, symptoms such as general loss of vitality, fatigue, and decreased potency and libido, which are not caused by testosterone deficiency, improve with mesterolone. In a double-blind, placebo-controlled study, 66 men with these symptoms received 75 mg mesterolone or placebo. Mesterolone brought about a significant improvement in symptoms compared with placebo[203]. Mesterolone also appears to alleviate mild depression[167]. It is not clear whether this effect is due to the psychotropic effects of androgens, which are well-known.

4.1.1.2. Oral testosterone undecanoate

In many countries and under various brand names, testosterone undecanoate (☞ Figure 4.7) is commercially available in 40 mg capsules (Organon). By esterification in the 17β-position with a medium-length fatty acid, when orally administered, testosterone undecanoate is not transported through the portal system to the liver but reaches the blood in chylomicrons through the lymphatics of the intestine and the thoracic duct. In this way it avoids, at least partially, the 'first-pass' effect of the liver and achieves a short-term increase in serum testosterone levels. However, oral testosterone undecanoate suffers from low bioavailability and a large inter- and intraindividual variability in absorption[116,344,360]. Uptake from the intestine is improved by dissolving testosterone undecanoate in oil, and taking it together with a fatty meal. A new capsule preparation has been developed which no longer requires storage in a refrigerator. Maximum serum testosterone levels are attained after an average of 5 h. Peak concentrations are between 17 and 96 nmol/l, reflecting the fluctuating bioavailability. However, testosterone levels fall significantly again after a few hours, and therefore 160-240 mg must be administered daily in three divided doses; even then there may be long periods when testosterone levels are reduced. Therefore, oral testosterone undecanoate is less suitable than parenteral testosterone for long-term substitution therapy. It is more suitable for patients who cannot tolerate transdermal or intramuscular administration. It can also be used to induce puberty, when initially only low testosterone levels are required.

4.1.1.3. Buccal testosterone

In the strive for a non-invasive parenteral testosterone delivery a buccal application (Striant®) has been developed. Striant® is a small tablet containing 30 mg testosterone and placed every 12 hours on the gingiva above the right or left incisor in the nasal fossa. As the tablet absorbs moisture it gradually dissolves and releases testosterone, which is absorbed across the oral mucosa. Thereby the first pass effect of the liver is circumvented and significant serum levels of testosterone can be achieved. After the first dose, serum testosterone levels peak after 10-12 hours. After the second application a steady state is reached and in most patients serum testosterone levels are well within the normal range. After removal of the tablet testosterone levels fall below normal after 2-4 hours. Striant® was evaluated in an unpublished multicenter, open label, single arm trial in 98 hypogonadal men. In this study, Striant® was administered twice daily for 12 weeks. The study was not designed to assess whether eating, drinking, toothbrushing or gum chewing influenced absorption of testosterone, but patients were permitted to engage in these activities during the trial. The mean daily serum testosterone concentration at week 12 was well within the normal range (520 ng/dl), however with a large variation (± 205 ng/dl). At week 12, over a 24 hour sampling period 76 % of the time serum testosterone levels were within the normal range. Serum

DHT and estradiol levels are raised in parallel with testosterone and remain in the normal range. 67 % of the patients found Striant® acceptable or even very acceptable, one third of the patients found it poorly or not acceptable. No information is available on clinical efficacy of Striant®. Other buccal application systems are under development[20].

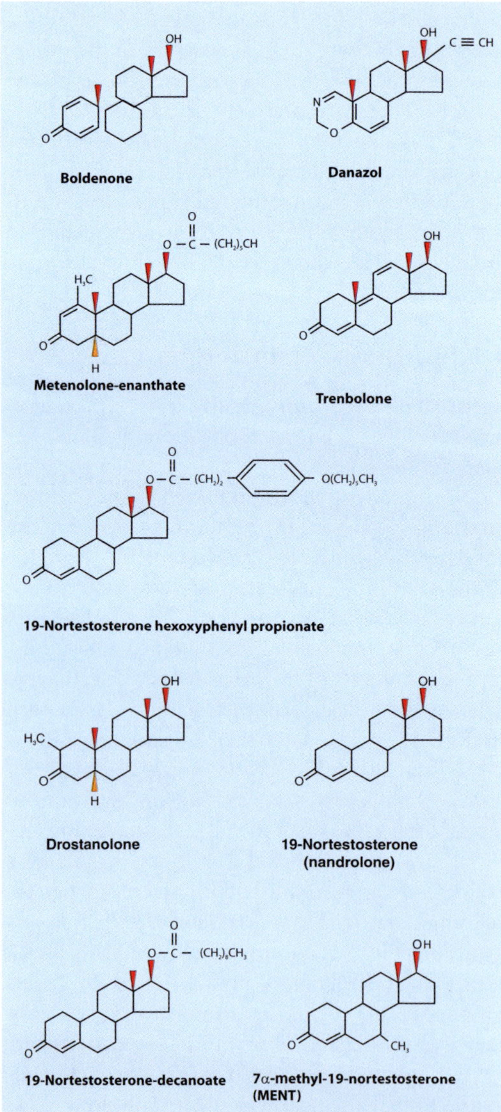

Figure 4.5: Androgenic steroids with stronger anabolic effect than testosterone and simultaneously weak androgenic effect (anabolics). 7α-Methyl-19-nortestosterone (MENT) will be the first androgen receptor modulator for clinical application.

4.1.1.4. 17α-alkylated androgens

Until the 1970s, 17α-alkylated androgens, in particular 17α-methyltestosterone, were used in substitution therapy for male hypogonadism, since they can be orally administered and the 17α-alkylation affords protection from rapid metabolism in the liver (☞ Figure 4.6).

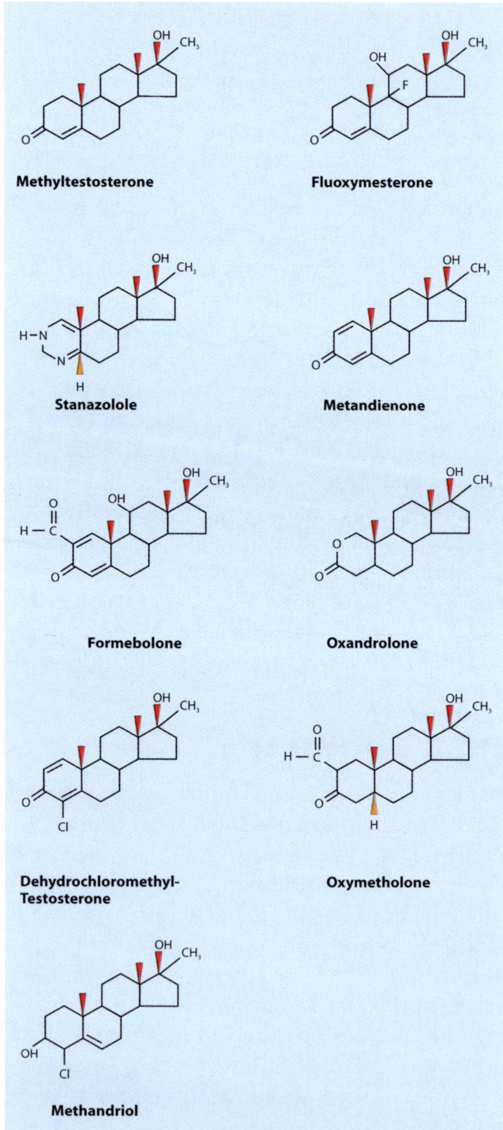

Figure 4.6: Composition of most important 17α-alkylated androgens which should no longer be used owing to their hepatotoxicity.

However, relatively high doses of 150 mg methyltestosterone a day are needed to obtain a clinical effect. Fluoxymesterone additionally carries a fluorine group in position 9 and a hydroxyl group in position 11, and is equally effective at low oral doses (20 mg per day). It has been proved, however, that all 17α-alkylated androgens are hepatotoxic, and may induce peliosis hepatis (hemorrhagic liver cysts) or jaundice, as well as benign and malignant tumors, regardless of dose[165]. For this reason these preparations were taken off the market in many countries. They are now considered to be obsolete. Unfortunately, intake of 17α-alkylated androgens is widespread in the context of anabolic misuse. People should be strongly advised against taking these substances (dehydrochloromethyltestosterone, fluoxymesterone, formebolone, methandienone, methyltestosterone, oxandrolone, oxymetholone, stanozolol).

4.1.2. Intramuscular and subcutaneous administration

4.1.2.1. Testosterone propionate

Testosterone propionate is a 17β-ester, which is very rapidly released from the depot injection owing to its very short side-chain. It therefore produces a high serum testosterone concentration very quickly (☞ Figure 4.7).

Intramuscular administration of 50 mg causes supraphysiological peak levels of testosterone after half a day. However, these fall below the normal range for men after only 2-3 days. Terminal half-life is only approximately 19 h. Long-term substitution with testosterone propionate necessitates at least two intramuscular injections per week. Therefore, for practical reasons, testosterone propionate is hardly ever used.

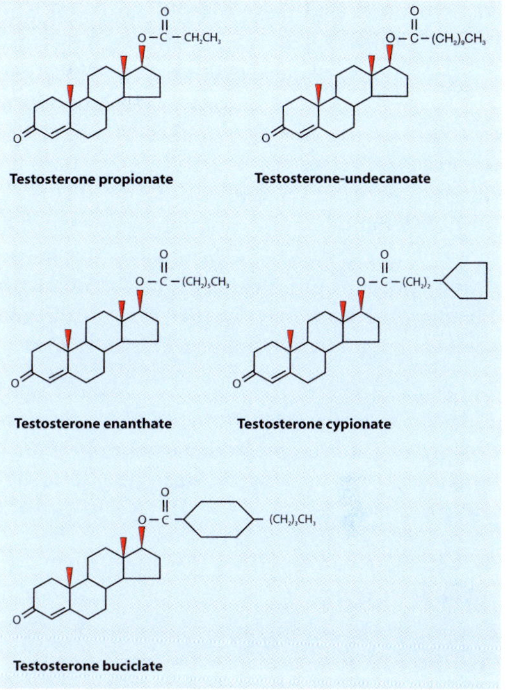

Figure 4.7: Structural formulae of 17β-esters of testosterone. Testosterone enanthate and undecanoate are commonly used substitution preparations in male hypogonadism. Testosterone buciclate is still under development.

4.1.2.2. Testosterone enanthate and testosterone cypionate

As the duration of effect of testosterone esters is dependent on the length of the esterified aliphatic side-chain in the 17β-position, testosterone enanthate has an essentially longer half-life and duration of effect compared with testosterone propionate (☞ Figure 4.7).

Intramuscular injections of 250 mg testosterone enanthate every 18-28 days provide adequate long-term substitution, although, 24-48 h after the injection, non-physiologically high serum testosterone concentrations of up to 80 nmol/l occur, followed by an exponential decrease to subnormal levels before the next injection[282,342,343]. After a 250-mg injection, serum testosterone levels fall below the normal limit for adult men after 2.5-3 weeks. Terminal half-life is 4.5 days. Therefore, injections of testosterone enanthate are usually given every 3 weeks and the administration interval adjusted immediately before the next injection, based on the trough level.

Many patients find these sharp fluctuations (initially significantly increased testosterone levels falling to hypogonadal levels by the next injection) to be unpleasant, since sexual function, mood and physical capacity are dependent on serum testosterone levels. Testosterone enanthate had been the standard preparation for testosterone treatment for decades and has proved to be safe in terms of unwanted side effects[178]. Nowadays, the longer acting injections of testosterone undecanoate (☞ Chapter 4.1.2.3.) seem to be preferred due to convenience and more physiological testosterone levels.

17β-ester testosterone cypionate and testosterone cyclohexane carboxylate have the same pharmacokinetic and pharmacodynamic profile as testosterone enanthate[342,343].

4.1.2.3. Intramuscular testosterone undecanoate

Intramuscular testosterone undecanoate (Nebido®) became first available for long-term treatment of hypogonadal men in Germany and Finland in 2004. The detailed studies on its pharmacokinetic behaviour revealed a significantly longer half-life and duration of effect than for testosterone enanthate[281,340].

Following a single administration of 1000 mg of testosterone undecanoate (Nebido®) to 14 hypogonadal men testosterone serum levels were raised into the normal range (12.3 nmol/l) within 2 days and mean maximum testosterone concentrations were 22.0 nmol/l at about 14 days after injection. Testosterone levels remained in the therapeutic range for 6-8 weeks. Subsequent studies showed that serum testosterone levels accumulated and exceeded the physiological range when Nebido® was administered in regular intervals of six weeks[281]. Therefore the injections intervals were gradually extended to 7-11 weeks followed by a regular injection interval of 12 weeks. Thus, intervals of around 3 months were found to maintain best testosterone serum levels within the normal range while avoiding non-physiological peaks. Therefore, only four intramuscular injections of testosterone undecanoate are required per year for long-term substitution[340] (☞ Figure 4.8). The terminal half-life of testosterone undecanoate after intramuscular injection is between 50 and 60 days. In terms of duration of effect, testosterone undecanoate injections lie between testosterone enanthate and implantable testosterone pellets. Nebido® combines the long duration of effect and steady active ingredient concentration of the pellets with the practicality of the testosterone enanthate injection[340,414].

It is recommended to reduce the injection interval between the first and second administration of testosterone undecanoate to 6 weeks. With this loading dose, sufficient steady state testosterone levels may be achieved more rapidly. After this initiation regimen further injections should be given in intervals of about 12 weeks. Measurements of serum trough testosterone levels in conjunction with clinical symptoms should be used to adjust the in-

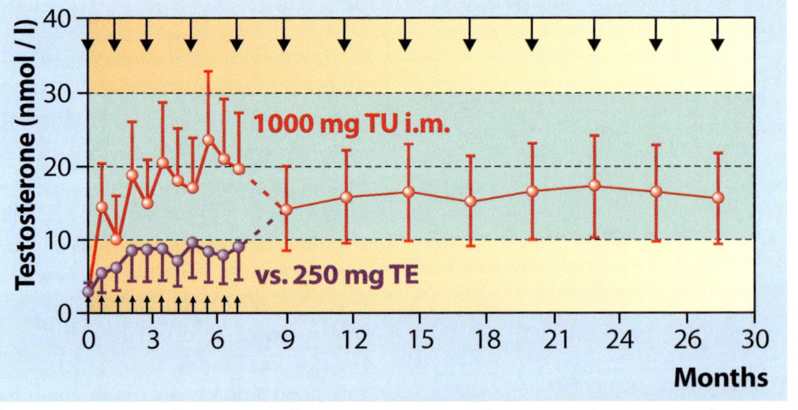

Figure 4.8: Pharmacokinetics of intramuscular testosterone undecanoate. In a comparison study, men with hypogonadism received either 250 mg testosterone enanthate (TE) or 1000 mg testosterone undecanoate (TU) intramuscularly every 3 weeks for the first 7 months (black arrows). The groups were subsequently combined and TU was injected every 6 weeks, then every 9 weeks and finally every 12 weeks[340].

jection interval individually. Serum trough levels should be in lower third of the normal range (12-18 nmol/l). Serum trough levels below this range would indicate a shortening of the injection interval by one week, and serum trough levels above this range suggest to extend the injection interval. In either case, a reevaluation of serum trough levels has to be made after 2 injections with the adjusted schedule.

Experiences of patients with regard to the following are very favorable: bone density, erythropoiesis, muscle mass, libido and potency. Therefore, intramuscular testosterone undecanoate has become the standard preparation for substitution in male hypogonadism. In particular, patients value the steady concentration of active ingredient, the consistent, reliable efficacy and the long duration of effect (☞ Figure 4.9). In comparison to conventional testosterone injections the use of Nebido® allows to avoid unphysiological high peaks and low troughs and to reduce drastically the frequency of injections, the most common complaint of patients on testosterone injections.

4.1.2.4. Testosterone implants (pellets)

Until the development of testosterone preparations which can be injected, the normal procedure for substitution was to implant testosterone pellets. Nowadays, subcutaneous implantation of testosterone is very popular in Great Britain and Australia. When it emerged that there were drawbacks associated with testosterone depot preparations, testosterone pellets underwent a revival[189] and their clinical efficacy was evaluated[179,180,190].

Testosterone pellets (Testosterone-Implant®, 100 mg or 200 mg, Organon) are licensed in Great Britain, Australia and some other countries. The pellets are made from pure crystals of testosterone and contain no excipients. Testosterone is pressed into cylindrical form at high temperature (154°C) and under high pressure. The 200 mg pellets measure 1.2 x 4.5 mm. Implantation is performed under sterile conditions and local anesthetic, after making a stab incision in the subcutaneous adipose tissue of the skin of the abdomen. Only 3-5 % of patients experience local side-effects in the form of infections, and these can be effectively treated with antibiotics.

Testosterone pellets have the longest duration of effect of all the substitution products available. Implantation of 6 x 200 mg pellets keeps serum testosterone levels within the normal range for approximately 6 months. Therefore, the majority of patients need only two implants per year. Immediately after implantation, supraphysiological testosterone levels occur for 1-2 days, but these are no higher than the peaks following injection with testo-

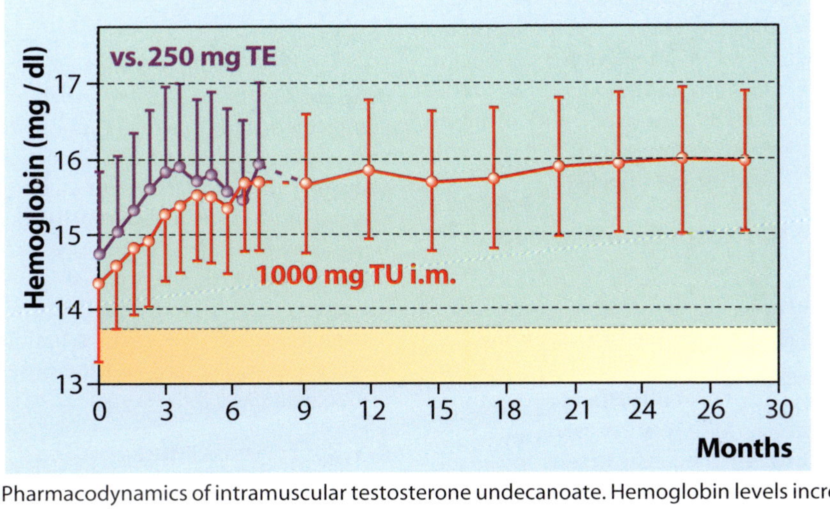

Figure 4.9: Pharmacodynamics of intramuscular testosterone undecanoate. Hemoglobin levels increase excessively over the 3-week period covered by the testosterone enanthate (TE) injection. This does not happen with the testosterone undecanoate (TU) injection. The non-physiologically high peak concentrations which occur following injection of TE cause excessive stimulation of erythropoiesis. This emphasizes the need for a formulation which has steady active ingredient levels[340].

sterone enanthate. After this, testosterone levels remain in the upper normal range for 2 months and then slowly return to lower normal limits (around 10 nmol/l) by the 6th month. Terminal half-life is 71 days, and far exceeds that of all other available preparations[189]. The availability of pellets containing different quantities of testosterone (100 mg and 200 mg), and the number of implantable pellets available, enables a wide variation in duration of effect.

Because testosterone pellets maintain a long and steady concentration of testosterone, they proved to be superior to oral and intramuscular testosterone in a randomized, comparative study on men with hypogonadism[179,180,189,190]. As the level of testosterone is constant, this obviously creates a stronger androgenic effect than with injectable esters. Hence, administering testosterone pellets to men with primary hypogonadism can suppress the gonadotropins to normal levels. This cannot be achieved with any other testosterone preparation. Research is currently being carried out to determine whether this will lead to greater benefits in the long-term, e.g. better stimulation of bone metabolism. Patients value pellets because they give stable testosterone levels and reliable and consistent physical and sexual function. Patients are also not reliant on frequent injections or tablets. Given a choice, patients clearly prefer implanted pellets over other substitution methods[189]. One fundamental disadvantage of pellets is the implantation procedure. This is labor-intensive for the doctor (approximately 15 min) and expensive, as conditions have to be sterile. The availability of intramuscular testosterone undecanoate (Nebido®) will considerably reduce the use of pellets.

4.1.2.5. 19-Nortestosterone and its esters

19-Nortestosterone, also known as nandrolone, is made by replacing the methyl group on the C10 atom by a hydrogen atom (☞ Figure 4.5). This steroid shows increased anabolic effects and reduced androgen effects. Whereas 19-nortestosterone binds more strongly than testosterone to the androgen receptor, 19-nordihydrotestosterone binds less strongly to the androgen receptor than does natural dihydrotestosterone. All tissues on which testosterone has a direct effect, e.g. musculature and bone marrow, therefore experience stronger stimulation, whereas tissues which are dependent on dihydrotestosterone (e.g. skin, prostate) are not stimulated as much. Accordingly, 19-nortestosterone does not have the same pharmacodynamic effects as testosterone.

These differentiated properties of a selective effect on various organs are currently being researched using the highly potent synthetic androgen 7α-methyl-19-nortestosterone (MENT, ☞ Figure 4.5)[375]. MENT is approximately ten times as potent as testosterone in bioassays and in negative feedback to the gonadotropins, but is only half as effective as testosterone in the prostate[88]. The selective tissue-specific effect is obtained because MENT has a stronger effect than testosterone, but cannot be reduced to dihydro-19-nortestosterone as the 7α-methyl group sterically hinders the enzyme 5α-reductase. Therefore, in bioassays it hardly stimulates the prostate at all. In contrast, MENT can be easily aromatized to estradiol[376] and should therefore maintain bone mass, although this has yet to be shown. MENT does not bind to the transport protein SHBG and is therefore rapidly eliminated from the circulation. MENT acetate, however, can be implanted in pellet form for long-term efficacy. Possibly a MENT gel might be developed. The few clinical studies that have been carried out show that MENT maintains male sexual activity and libido, although it severely suppresses gonadotropins and endogenous testosterone[16].

Hence, MENT could become the first selective androgen receptor modulator, whereby the androgen receptor is not actually modulated. MENT's unique property could mean a beneficial androgen effect on the musculature, bones, psyche (e.g. libido) and other target tissues which depend on testosterone, while at the same time avoiding potentially adverse effects on the prostate (growth, cell proliferation) and skin (e.g. acne). All this makes evaluation of MENT for use in the inhibition of male fertility (contraception) and late-onset hypogonadism particularly interesting.

4.1.3. Transdermal application

4.1.3.1. Testosterone gel

The latest transdermal formulation to be developed is testosterone gel, which is available as Testogel® and/or Androgel® in the US since the year 2000 and in Europe since 2003. Later a second testosterone gel under the brandname Testim® became available, which uses the same dosing regimes. This is a clear anhydrous alcohol gel containing 1 % testosterone. It is applied once a day to the abdomen, shoulders or upper arms. Testosterone gel has excellent pharmacokinetic properties and is very well tolerated. Approximately 10 % of a 5 g dose of the gel, equivalent to 50 mg testosterone, is absorbed and reaches the systemic circulation. Because application is through the skin, the first-pass effect of the liver is avoided, and therefore one 5-10 g dose of gel per day can totally replace endogenous testicular production.

Thirty minutes after applying the gel, blood testosterone levels start to rise. After 2-4 h they are within the normal range. The gel is absorbed very rapidly by the skin. The cutis acts as a reservoir and releases testosterone constantly into the systemic circulation for 24 h so that it does not fall below normal limits. After 2-3 applications the system is saturated, and there is an equilibrium between absorption and elimination (☞ Figure 4.10). The extent of surface area to which the gel is applied is of minor importance. Testosterone levels were only 20-30 % higher when the surface area was quadrupled in size[417].

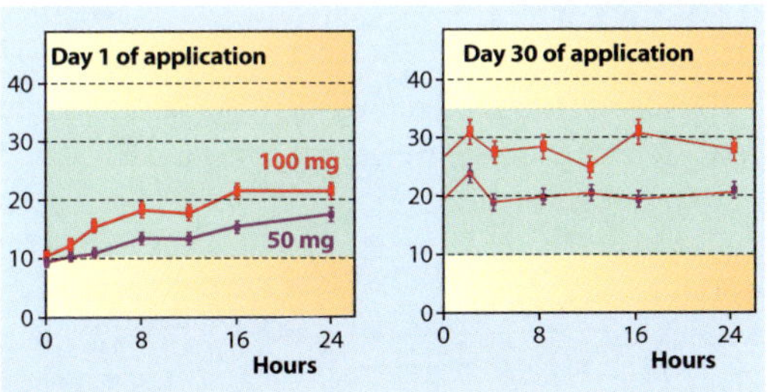

Figure 4.10: Daily testosterone profile in men with hypogonadism on day 1 and day 30 of gel application. Serum testosterone levels remain within the normal range after 2-3 doses. Even peak and trough levels stay within normal limits[380].

Continuous release from the skin prevents severe fluctuations in testosterone levels, which then remain within normal physiological levels if the gel is continuously applied. Levels seldom exceed peak physiological levels or fall below minimum levels. The 5 g dose is suitable for 60 % of patients. Approximately one-third of patients require a higher dose (7.5 g per day, rarely 10 g). In most countries, testosterone gel is available in two different dose sizes: the 2.5 g gel containing 25 mg testosterone and 5.0 g gel containing 50 mg testosterone. Therapy should be started with the 5.0 g gel and adjusted as necessary to 10.0 g. Once testosterone substitution is finished, original endogenous testosterone levels are attained 3 days after the final application of gel, owing to the reservoir effect of the skin.

Regular application of gel normalizes serum DHT and estradiol concentrations without causing high peaks of these metabolites. LH and FSH are suppressed in patients with primary hypogonadism, which indicates good bioavailability. In an extensive multicenter study carried out in men with hypogonadism, doses of 5-10 g testosterone gel per day were more effective (in terms of increase in lean body mass, bone density, sexual satisfaction and decrease in fat mass) than testosterone patches. Muscle strength and hematopoiesis increased in all groups[420,421]. In men over 60 years of age with reduced testosterone levels, bone density of the spine increased by 0.9 %, lean body mass increased by 3.9 % and fat mass fell by 4.4 %.

Testosterone gel is very well tolerated by the skin. In a randomized, controlled study in 227 men with hypogonadism, 5.5 % of the men had mild skin irritations at the application site after 3 months of applying the gel, compared with 66 % of men in the parallel testosterone patch group[380]. In the 30 months followup of the initial studies, and during the first year after launch in the USA, no significant side-effects have been reported except for rare cases of skin irritation.

Since the gel contains alcohol it should not be applied to mucous membranes or come into contact with the eyes. Genital skin is also an unsuitable site for applying the gel, since the alcohol in the gel can cause irritation. It also contains a high concentration of the enzyme 5α-reductase, which would metabolize an excessive amount of testosterone to DHT if it penetrated the skin. During the first few hours after applying the gel, close skin contact can cause testosterone to transfer to a partner. A longer interval (at least 6 h) significantly reduces the transfer rate, and washing the skin with soap or covering oneself with a shirt completely prevents transfer of testosterone.

In general, results from studies of testosterone gel are excellent. The gel has the best pharmacokinetic properties of all the available formulations. This is associated with very good efficacy and only rare cases of skin irritation. Hence, there is a higher compliance rate among patients using this formulation. In a randomized, controlled study, compliance among users of testosterone gel was 93-96 %, but was only 65 % among users of patches[420].

4.1.3.2. Dihydrotestosterone gel

A gel containing dihydrotestosterone (2.5 %) is on the market in France and Uruguay. Doses of 5 mg containing 12.5 mg of DHT are applied to a large area of the trunk (Andractim®, Besins-Iscovesco, Paris). After single daily application, significantly increased levels of DHT are measured in the serum already within the first day. Testosterone and 17β-estradiol do not increase, as DHT cannot be converted into these steroids. Accordingly, this formulation does not display the full testosterone effect. Whilst the general clinical effect is satisfactory[332], owing to a lack of data it is not clear whether osteoporosis can be corrected by the administration of DHT. As this effect is mediated by 17β-estradiol, an osteoanabolic effect is not anticipated. Therefore, percutaneous therapy with DHT is not suitable for substitution of male hypogonadism and is reserved for special indications, e.g. for topical application where the penis is abnormally small (micropenis). Other gels containing DHT which have a more favorable pharmacokinetic profile are currently in development[419].

4.1.3.3. Scrotal testosterone patches

Between 1998 and 2001, scrotal testosterone patches which had already been licensed for some time in the USA became available in Europe under the brand name Testoderm® 15. The patches contain 15 mg testosterone on a gauze dressing. When in contact with the skin, approximately 6 mg is released over 24 h. This corresponds very closely to the physiological rate of testosterone production of healthy testes. Nevertheless, this rate of release can only be attained if applied to the scrotum, and the scrotum has to be shaved so that the patch can stick properly. The skin and subcutis of the scrotum are very thin, and sufficient quantities of testosterone can be absorbed from the patch only if it is applied to the scrotum. Testoderm® 15 is ineffective if applied elsewhere.

Maximum serum levels are attained by about 2-4 h after application. These fall again to initial values within 22-24 h. Accordingly, a new patch must be applied every day. Maximum testosterone levels are achieved in the morning if the patch is applied late evening or very early in the morning. This mimics the physiological circadian rhythm very well. In contrast to some injectable testosterone esters, there are no non-physiologically high testosterone levels[22,90]. Apart from the steady testosterone levels, many patients value not having to be dependent on intramuscular injections of testosterone esters, which can also be painful. However, patches require the co-operation of the patient, who has to apply them daily and shave his scrotum in advance. About 5-8 % of patients complain of local itching. Testoderm® is not suitable for patients with a very small scrotum, e.g. in Klinefelter's syndrome or anorchidism.

Since scrotal skin harbors high activities of the enzyme 5α-reductase, a large amount of testosterone is converted to 5α-dihydrotestosterone during penetration of the skin. Therefore serum DHT concentrations permanently exceed normal ranges by 4-5-fold[5,90] (☞ Figure 4.11). The shift in the ratio of DHT/testosterone is being critically evaluated, as DHT might be involved in the development of hyperplasia of the prostate, and the long-term effect of a pathologically increased DHT/testosterone ratio is totally unknown. However, no adverse effects have been reported to date. Scrotal testosterone patches are very expensive, compared with other substitution preparations.

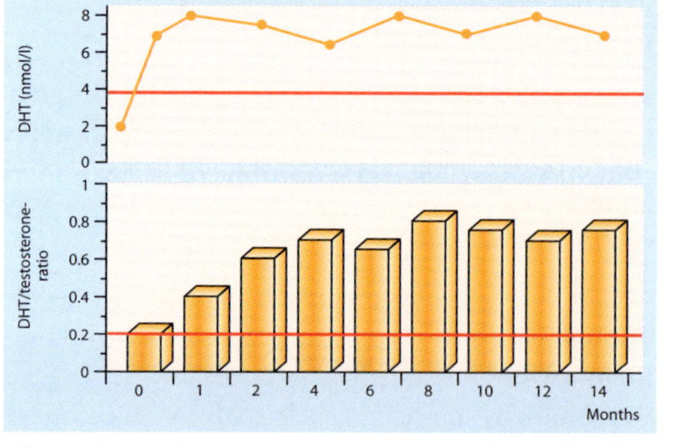

Figure 4.11: Serum concentrations of 5α-dihydrotestosterone (DHT) and ratio of DHT to testosterone during long-term therapy with scrotal patches. The horizontal lines show the upper normal limit[22,90].

4.1.3.4. Non-scrotal testosterone patches

Second-generation testosterone patches have a reservoir containing gel in which 12.2 mg testosterone is dissolved. Addition of an absorption amplifier enables 2.5 mg to be released over 24 h through non-scrotal skin. Androderm® 2.5 mg is applied to the shoulders, upper arms, hips or abdomen, each day to a different site. As the patches are applied to non-scrotal skin, the DHT/testosterone ratio does not shift outside the physiological range[256,258]. Peak levels are attained after approximately 8 h and may slightly exceed the normal limit, but they fall again to the lower normal limit 24 h after application, and therefore a new patch has to be applied each day. The majority of patients have to wear two patches simultaneously to attain an adequate level of active ingredient. Only a few patients manage on one patch a day. At least 9 % of patients have to discontinue therapy with Androderm® 2.5 mg because of skin irritation and contact dermatitis[199,257]. In one study almost 2 thirds of the patients suffered from skin reactions[380,420]. The benefits of Androderm® 2.5 mg over scrotal patches include that they can be applied to non-scrotal skin and that treatment costs are somewhat lower. Both patches have favorable pharmacokinetic properties which mimic the circadian rhythm.

4.1.4. Future and experimental possibilities of androgen substitution

The unfavorable pharmacokinetic properties of the most commonly used testosterone esters has led to the development of new testosterone esters and new formulations of testosterone since the 1980s. The spectrum of treatments available for male hypogonadism now includes buccal testosterone, testosterone gel, testosterone patches and an intramuscular formulation of testosterone undecanoate (Nebido®).

Clinical studies have shown the pharmacokinetic behavior of testosterone microspheres for intramuscular injection to be very similar to that of testosterone pellets[43]. The microcapsules contain testosterone which is (micro)encapsulated in biologically degradable lactide-glycolide copolymers. When the copolymers dissolve, testosterone is released gradually, and therefore there is a reservoir effect as with testosterone pellets. Accordingly, testosterone microspheres show zero-order release kinetics with a serum testosterone plateau phase of 20 nmol/l over 2 months. After the plateau phase there is a rapid decrease into the hypogonadal range, as the half-life of 19 days indicates. The testosterone effect lasts for approximately 2.5 months. The requirement for two simultaneous intramuscular injections of 2.5 ml is still a disadvantage. In addition, the suspension is very viscous, and tends to block the needle during injection, so that often more than two attempts at injection must be made[43].

Another new development is the testosterone-hydroxypropyl-β-cyclodextrin complex (TCD), in which pure testosterone is surrounded by a carbohydrate ring which promotes absorption through the mucous membrane of the mouth (☞ Figure 4.12). Sublingual administration of 2.5-5 mg TCD leads to peak serum testosterone levels of between 60 and 80 nmol/l after only 25 min[373,416]. Initial values are attained after only 2 h, and therefore it has to be taken several times a day. The bitter taste and short duration of effect has so far limited the use of TCD to experimental administration. Its development has been discontinued.

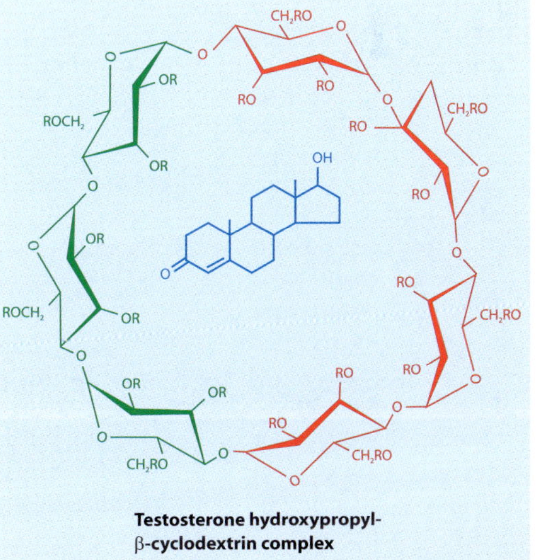

Figure 4.12: Structure of testosterone-hydroxypropyl-β-cyclodextrin complex carbohydrate ring containing testosterone molecule.

A testosterone-17β-ester, testosterone buciclate, which has been developed by the WHO, has a very long half-life and duration of effect compared with other testosterone esters. This is due to the long aliphatic side-chain and ring structure (☞ Figure 4.7). Unlike other testosterone esters, testosterone buciclate is injected not in an oily solution but micronized in an aqueous solution. A single intramuscular injection of 600 mg testosterone buciclate can raise the serum testosterone level by approximately 6 nmol/l[31]. Testosterone buciclate does not cause supraphysiological peak levels immediately after injection, and therefore serum testosterone levels remain constant. Testosterone concentrations in the saliva, an index of free, biologically active testosterone, remain within the normal range for 7 weeks and then fall rapidly. Terminal half-life is 29.5 days and therefore significantly longer than that of other testosterone esters, but considerably shorter than those of testosterone pellets and intramuscular testosterone undecanoate. The development of testosterone buciclate has been stopped.

4.1.5. Summarizing assessment and differential therapy

The most important indication for the administration of androgens is testosterone deficiency in the context of male hypogonadism. The aim of substitution is to replace the missing testosterone and establish normal levels of circulating testosterone appropriate to a person's age.

> As the spectrum of effect of endogenous testosterone is essentially based on its metabolism to dihydrotestosterone and estradiol, only testosterone can be used to treat male hypogonadism.

An ideal testosterone preparation should provide steady testosterone levels which fluctuate within the range of the physiological circadian rhythm and simultaneously requires only few applications. Testosterone gel and the testosterone patch come very close to this ideal, with serum testosterone levels remaining almost exclusively within the physiological range. The best physiological testosterone levels have been attained with the gel. This is reflected in very good efficacy. Currently, the non-oral testosterone therapy with the fewest side-effects is Testogel® (☞ Table 4.3, Figure 4.13).

Administration	Active ingredient	Formulation	Terminal half-life	Duration of effect
Oral	testosterone undecanoate	40 mg capsules	1.6 h	3-4 h
	buccal testosterone	30 mg tablets	2 h	6-10 h
Intramuscular	testosterone propionate	50 mg ampoules	0.8 h	2.4 days
	testosterone enanthate	250 mg ampoules	4.5 days	12 days
	testosterone microspheres	not commercially available	18.8 days	77 days
	testosterone buciclate	not commercially available	29.5 days	84 days
	testosterone undecanoate	1000 mg ampoules	approximately 55 days	approximately 90 days
Subcutaneous	testosterone pellets	100 mg and 200 mg pellets	70.8 days	180 days
Transdermal	testosterone	gel containing 50 mg T	1 day	24 h
		gel containing 25 mg T		
		12.2 mg patch	1 day	24 h

Table 4.3: Common testosterone preparations.

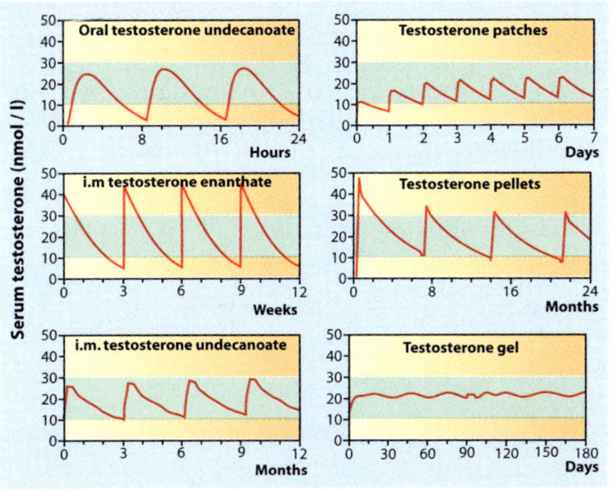

Figure 4.13: Schematic overview of serum testosterone levels after administration of selected testosterone preparations. The normal range is indicated. An oral dose of 40 mg testosterone (T) undecanoate every 8 h causes a short-term increase in serum testosterone to within the normal range. However, this falls again rapidly to subnormal levels. Daily application of testosterone patches increases serum testosterone well inside the normal range, and closely mimics the physiological circadian rhythm. Intramuscular (i.m.) administration of 250 mg testosterone enanthate every 21 days causes supraphysiological levels of testosterone initially, and later subnormal levels. For approximately one-third of the time between injections, the patient is under- or overdosed. Subcutaneous implantation of six 200 mg testosterone pellets causes mild supraphysiological peak levels, which then slowly decrease over a period of months and guarantee long-term substitution. Intramuscular testosterone undecanoate, causes very steady levels over a lengthy period when administered in a dose of 1000 mg every 12 weeks. The peaks and troughs remain within the normal range. Testosterone gel is characterized by very stable and uniform serum testosterone levels. Daily application does not cause significant fluctuations.

The disadvantage of scrotal patches is that they produce non-physiologically high DHT levels, and the long-term consequences of this are not clear. Moreover, regular shaving of the scrotum to enable the patches to stick requires compliance. The non-scrotal patch does not have these disadvantages, although it is associated with a high incidence of skin irritations. In addition, it can be felt and heard (rustling), which does not help compliance. Therefore, testosterone gel with its good efficacy and pharmacokinetics will rapidly become one of the most popular transdermal testosterone preparations, especially as the standard preparation for substitution in male hypogonadism to date, the 3-weekly injection of testosterone enanthate, is not ideal. The unfavorable pharmacokinetic profile with high serum peaks and troughs before the next injection are often considered unpleasant by patients. Considerably steadier active ingredient levels are attained with the testosterone undecanoate injection ('3-monthly injection'). This is associated with stable testosterone levels which remain permanently within normal ranges, good clinical efficacy and very high patient compliance. Although testosterone pellets have similar properties, they present difficulties, as implantation is costly and the infection rate associated with implantation is high. Therefore, injectable testosterone undecanoate will become the standard preparation for long-term substitution in male hypogonadism, in preference to testosterone enanthate and testosterone pellets.

The bioavailability of oral testosterone undecanoate is very variable, and it is therefore less suitable for long-term substitution in men with hypogonadism (☞ Table 4.4, Figure 4.13). It represents a sensible alternative where there is intolerance to the transdermal formulations of testosterone (gel, patches) or contraindications to intramuscular injections (e.g. simultaneous therapy with anticoagulants, hemodialysis patients). It has a proven efficacy in the treatment of delayed puberty, where a full substitution dose is not required during the initial stage. Hence, adequate levels are attained even with oral testosterone undecanoate. Testosterone enanthate is frequently used in this indication, and the dose can be individually tailored by giving a partial dose from one ampoule. It is significantly cheaper and, unlike oral capsules, not reliant on the compliance of the patient. Testosterone enanthate is almost exclusively used in androgen therapy of excessive growth because of the extremely high doses required. Experience with testosterone enanthate and oral testosterone undecanoate in the therapy of female-to-male transsexuality is well documented. Since these patients require total substitution after their ovaries are removed, oral testosterone undecanoate appears to be less suitable because of its restricted and variable bioavailability, although there are no direct comparative studies available. Recently, the first clinical experiences with the use of intramuscular testosterone undecanoate (Nebido®) in female-to-male transsexuals were reported and showed very favourable results.

Primary and secondary (classical) hypogonadism

- Intramuscular testosterone undecanoate ('3-monthly injection')
  - advantages: stable testosterone levels within normal range over months, high patient acceptance, no problems with compliance, very good efficacy
  - disadvantages: intramuscular injections (although only four per year), available in 2004
- Testosterone gel
  - advantages: very good pharmacokinetics, no fluctuation in testosterone levels, good efficacy, suitable if contraindications to injections, few side-effects, individual dosing, discrete application
  - disadvantages: inadequate long-term experience with the new product
- Testosterone enanthate ('3-weekly injection')
  - advantages: good track record, inexpensive, hardly any problems with compliance
  - disadvantage: frequent intramuscular injections, moderate pharmacokinetics with unpleasant fluctuations for patients
- Testosterone patches
  - advantages: good pharmacokinetics with uniform levels, suitable where contraindications to injections
  - disadvantages: scrotal patches: DHT increase, very expensive, non-scrotal patches: frequent, pronounced skin irritations, expensive
- Oral testosterone undecanoate
  - advantages: for patients who do not tolerate patches and where intramuscular injections are not possible
  - disadvantages: poor bioavailability, very expensive
- Testosterone pellets
  - advantages: good pharmacokinetics and pharmacodynamics, longest duration of effect of all preparations, high patient acceptance
  - disadvantages: costly to implant, local infections

Delayed puberty

- Testosterone enanthate
  - advantages: proven efficacy in this indication, doses can be tailored to individual requirements
  - disadvantages: intramuscular injections, poor pharmacokinetics
- Oral testosterone undecanoate
  - advantages: adequate levels for this indication, oral administration possible
  - disadvantages: administration several times a day requires good compliance, poor bioavailability, very expensive

Late-onset hypogonadism

- Testosterone gel
  - advantages: simple application, good pharmacokinetics with uniform levels, suitable where contraindications to injections, if required treatment can be stopped immediately (e.g. suspect prostate findings), efficacy in this indication shown in one study, few side-effects, individual dosing, discrete application
  - disadvantages: little experience with this indication
- Testosterone patches
  - advantages: good pharmacokinetics with uniform levels, suitable where contraindications to injections; if required, treatment can be stopped immediately (e.g. suspect prostate findings)
  - disadvantages: efficacy in this indication not supported by studies, therapy very expensive; scrotal patches: DHT increase; non-scrotal patches: skin irritations
- Testosterone enanthate
  - advantages: proven indication in studies, inexpensive
  - disadvantages: intramuscular injections, moderate pharmacokinetics with unpleasant fluctuations for patients, cannot be immediately withdrawn if required

Table 4.4: Differential therapy with androgens for the most important indications. DHT = 5α-dihydrotestosterone.

Mesterolone and anabolic steroids do not show the complete spectrum of effect of testosterone, and are not suitable in the indication of hypogonadism. The anabolics clostebol, metenolone and 19-nortestosterone decanoate are rarely used today, and the indication is almost exclusively restricted to consumptive diseases. Anabolics have been superseded in the treatment of osteoporosis by other treatments, particularly since the efficacy of anabolics has not been persuasively documented.

4.2. Indications for androgen administration

A distinction is made between androgen administration to compensate for androgen deficiency in primary or secondary hypogonadism and androgen administration for other reasons. Whereas compensation for a deficit is known as substitution, administration of androgens in other disorders is known as androgen therapy. The distinction between substitution and therapy is significant in that compensation of a deficit represents the indication *per se*, and benefit to the patient can be guaranteed. Therefore, both forms of hypogonadism represent an indication to correct the androgen deficiency if serum testosterone is low for the patient's age group (☞ Table 4.5), provided that there are no contraindications.

Proven indication
• primary and secondary hypogonadism in men
• induction of puberty in boys
• excessive constitutional growth in boys
• transsexuality (female → male)
Probable indication
• late-onset hypogonadism
• primary and secondary osteoporosis in men (to supplement osteoporosis medication)
• weight loss in consumptive diseases
• hereditary angioedema
• contraception in men (future)
Unproven indication
• aplastic and renal anemia
• abnormally small penis in newborn (micropenis)
Definitely not indicated
• to improve performance in sport
• erectile dysfunction without androgen deficiency
• idiopathic infertility

Table 4.5: Indications for androgen substitution.

4.2.1. Primary and secondary hypogonadism

Primary and secondary forms of hypogonadism are the classical indications for testosterone substitution (☞ Chapter 2.2.). After securing a diagnosis and repeatedly obtaining low serum testosterone levels, the androgen deficiency should be compensated, provided that there are no absolute contraindications to androgen substitution. Apart from correction of current symptoms of androgen deficiency (e.g. inadequate virilization, loss of libido and potency, muscle atrophy, androgen deficiency anemia), the aim of substitution is to prevent long-term sequelae of hypogonadism, e.g. osteoporosis.

The biological effects of testosterone develop slowly, and are therefore noticeable only several months after starting substitution therapy. Patients frequently report an increase or restoration of libido or potency as the first sign. Most men report normal libido and erections after only 3-5 months of testosterone substitution therapy. In particular, spontaneous morning erections return, which are often absent in testosterone deficiency[66,70]. Occasionally patients complain that erections remain inadequate, despite normalization of testosterone levels. Then, additional causes of erectile dysfunction should be sought. Often these patients suffer psychological problems and fear of failure as a result of the previous phase of impotence. In this case, transient administration of preparations that promote erection can restore self-confidence. Testosterone stimulates secretion in the seminal vesicles and prostate, and therefore volume of ejaculate increases too. This improves orgasm and satisfaction with sex-life as a whole.

Many patients feel that testosterone substitution has a beneficial effect on emotional state, self-confidence and activity. In studies of testosterone substitution in male hypogonadism, this aspect of the androgen effect was unfortunately neglected, and it is also rarely documented, not least because of problems with methodology. However, more recent studies clearly show that mood, social behavior and general activity are very positively affected and depression, anxiety and lethargy decrease[7,15,16,416,420].

Furthermore, two recent studies with men with refractory depression clearly show a major improvement when these men are treated with testosterone[76,308].

In addition, testosterone substitution was clearly shown to have a beneficial effect on musculature[45]. Within 6-12 months of starting therapy, there is a significant increase in overall muscle mass owing to an increase in the size of individual muscle cells, even though the number of cells stays the same. This results in a functional increase in muscle strength and capacity[42,60]. Overall, relative body composition changes and there is a decrease in fat tissue and increase in lean body tissue. Men with hypogonadism show a relative increase in fat mass. This is normalized by testosterone substitution. Epidemiological studies showed that testosterone levels are reduced in men with visceral obesity[24,347]. Testosterone levels inversely correlate with mass of visceral adipose tissue and plasma HDL cholesterol levels. Testosterone substitution in middle-aged men with obesity reduces insulin resistance, blood sugar and blood pressure[243,247]. Testosterone is an important factor in regional fat distribution in men[242]. Body weight regularly increases by 1-2 kg, usually as a result of water retention.

Patients with hypogonadism frequently suffer from osteopenia or manifest osteoporosis (☞ Figure 4.14).

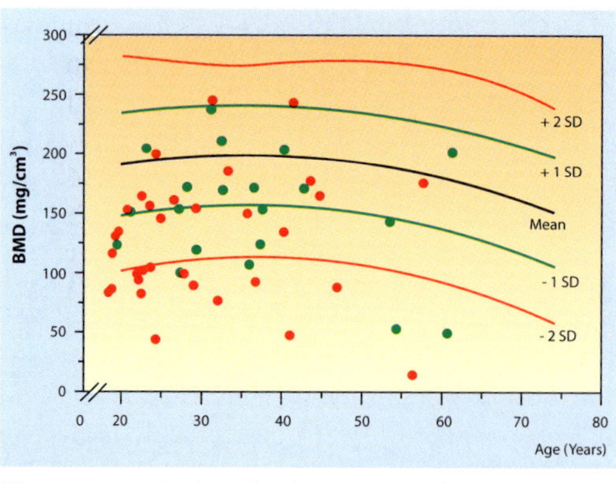

Figure 4.14: Trabecular bone mass of the ultradistal radius in 53 men with primary (green, n = 20) or secondary (red, n = 33) hypogonadism; 60 % of patients have osteopenia with bone mass that is reduced by more than 1 standard deviation (SD) and 36 % have osteoporosis (-2 SDs).

Bone mass is promoted by testosterone substitution, and is correlated to a patient's age and the extent of reduction in bone density. Younger patients respond better to testosterone administration than older men, and the lower the initial bone density the greater is the increase in bone mass[30,111,115,341]. Recent studies indicate that maintenance of bone density is closely correlated to the dose of testosterone administered[402]. Since the bone-anabolic effect is largely caused by aromatization of testosterone to estradiol, testosterone substitution must be given. For this reason, non-aromatizable androgens are not suitable for treating hypogonadism and osteoporosis.

The skin is an important target organ of androgens. Sebum production ceases when there is androgen deficiency, so the skin becomes dry and sensitive. Secondary male hair does not grow, or grows only slowly. Testosterone substitution rapidly causes an increase in sebaceous gland activity. Therefore, pimples or mild acne may initially appear in areas where there is this tendency (shoulders, chest, buttocks, chin). In rare cases, the dose may need to be reduced or treatment withdrawn for a while. The scalp and hair also become significantly more greasy and hair must be washed more often.

Sebum production increases relatively quickly after starting substitution therapy. It takes longer for hair to grow back - sometimes as long as 6 months. The testosterone effect is first evident in the armpit and pubic area. Some patients on substitution treatment slowly lose hair on the temples. This is called male alopecia ('receding hairline') and is a sign of virilization. It is a part of the testosterone effect. The patient should be told that this is a masculine attribute and not due to illness.

Before substitution, the prostate and seminal vesicles are frequently hypotrophic and appear small on sonography, owing to the deficit of testosterone. During substitution, the prostate returns to a normal size appropriate to the patient's age[29]. Increased growth is not anticipated at normal substitution doses. Benign hyperplasia or, more rarely, carcinoma of the prostate, can occur in older men on testosterone substitution. Therefore, recommendations for prostate screening also apply to men on androgen substitution[171].

The penis of an adult male rarely increases in size on testosterone substitution, and the testes shrink as a result of the negative feedback which testosterone and its metabolite estradiol exert on the gonadotropins.

Androgen deficiency anemia usually responds well to testosterone substitution, and the red cell count increases within a few months. This effect is dose-dependent and correlates with the serum testosterone level[190]. Normalization of serum testosterone causes an average increase in hemoglobin of around 2 g/dl, and in hematocrit of approximately 6 % (☞ Figure 4.15).

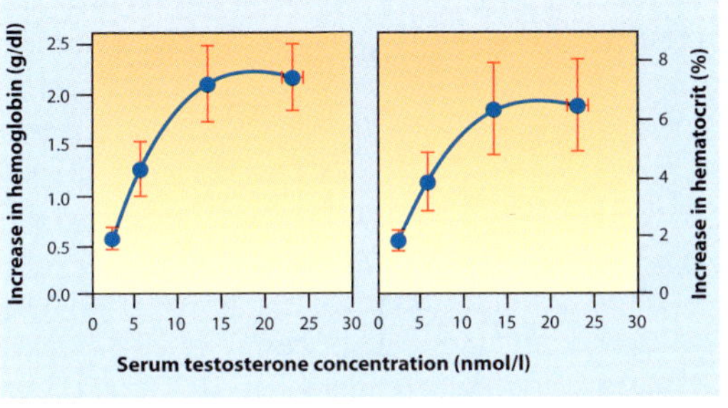

Figure 4.15: Dose-dependent stimulation of erythropoiesis by testosterone. A dose-related increase in hemoglobin and hematocrit was obtained as a factor of the serum testosterone levels attained by different androgens in men with hypogonadism[190].

If serum red cell count, hematocrit and hemoglobin do not increase, other causes of anemia must be sought. Iron deficiency is frequently encountered. If there is an excessive increase in hematocrit above the upper limit of normal, the dose of testosterone should be reduced or the interval between doses extended. Stimulation of erythropoiesis in particular appears to be dependent on peak substitution levels. It is therefore not as pronounced with a uniform substitution formulation (intramuscular undecanoate) as with formulations which produce very fluctuating levels (e.g. testosterone enanthate)[340].

Testosterone substitution usually causes a change in lipid metabolism[180]. During the pre-existing phase of hypogonadism, serum HDL cholesterol levels are often high. Testosterone causes a slight decrease in HDL cholesterol and occasionally increases LDL cholesterol. However, not all studies have convincingly shown this effect. Especially when substituting elderly men, several studies have shown beneficial effects of testosterone on lipid metabolism[420,441]. Whereas high doses of anabolics or testosterone substitution overdoses cause adverse changes in lipid metabolism, normal testosterone substitution doses do not increase cardiovascular risk. Androgens such as mesterolone, which do not aromatize, have the same effect as testosterone and its esters[180]. As substitution therapy in hypogonadism is aimed at compensating for a testosterone deficiency, if LDL cholesterol increases above the upper limit of normal (150 mg/dl), testosterone should not be withdrawn. Instead, one should probably consider administering a lipid-lowering drug (e.g. a cholesterol-synthesis inhibitor). On the other hand, if there is a pronounced reduction in HDL cholesterol during testosterone substitution, one should consider reducing the dose, taking into account additional parameters (hemoglobin concentration, hematocrit, serum testosterone). Intramuscular testosterone undecanoate and testosterone gel showed no significant effect on HDL and LDL cholesterol in clinical studies[340,420]. However, even in the clinical studies there was a tendency for HDL cholesterol to decrease.

Testosterone substitution does not normally promote fertility. If men with secondary hypogonadism want to have children, pulsatile GnRH or hCG/hMG treatment can stimulate the germinal epithelium. Men with primary hypogonadism wishing to have children should be referred for assisted fertilization. As short-term testosterone substitution does not adversely affect later stimulation of the germinal epithelium by GnRH or gonadotropins, testosterone treatment is always administered initially, even to men with secondary hypogonadism; this can later be temporarily replaced by GnRH or hCG/hMG therapy (☞ Figure 4.16). After successful conception, testosterone substitution treatment is re-instated, since GnRH is essentially more costly and more labor-intensive to administer[198,435].

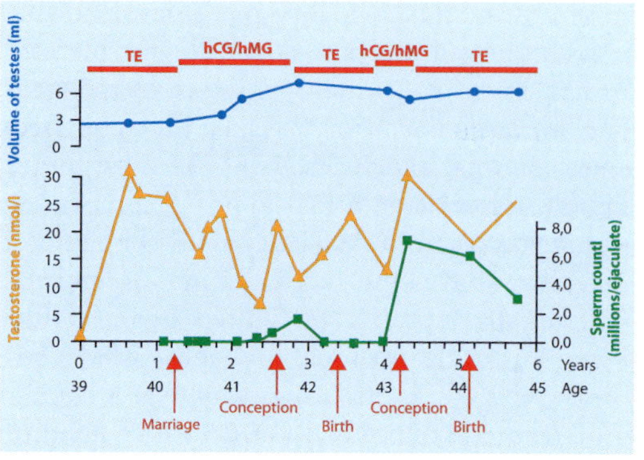

Figure 4.16: A patient who was not diagnosed with Kallmann's syndrome until the age of 39 years was initially given testosterone enanthate (TE) substitution treatment. This normalized the low serum testosterone levels (yellow) and increasing virilization was achieved. After over a year of testosterone substitution therapy, the patient married and expressed a desire to have children. His therapy was changed to human chorionic gonadotropin/human menopausal gonadotropin (hCG/hMG) to stimulate spermatogenesis. This stimulated the germinal epithelium resulting in growth of the testes (blue). Concentration of sperm in the semen rose during therapy from 0 to a maximum of 4.3 x 10^6/ml (green). His wife conceived after 15 months. Then he was changed back to TE therapy. When he wanted another child, he was again given hCG/hMG, which resulted in pregnancy after only 6 months. Since then he has been substituted only with testosterone.

4.2.2. Induction of puberty

One should always consider inducing puberty in young boys with delayed puberty and psychological problems owing to delayed development. After adequate diagnosis (☞ Chapter 2.2.1.2), it is immaterial to therapy whether the patient is suffering from constitutional delay of development (CDD, 'late developer') with a slow biological clock or primary or secondary hypogonadism with permanent testosterone deficiency. If delayed puberty is due to other factors (e.g. severe general (systemic) illness, mass in the hypothalamic or hypophyseal region) an attempt should be made to correct the underlying condition. If androgen deficiency persists, these boys should then be treated with testosterone. If the dose is titrated correctly for age, there should be no risk of premature closing of the epiphyses and reduced final height[9]. Hence, induction of puberty should not be delayed until late adolescence. The aim of therapy is to bring the puberty of the patient into line with that of his peers. Not only does this involve virilization at the appropriate time, but also stimulation of growth in height. In the early stage of puberty, testosterone stimulates production of growth hormone, initiating growth appropriate to chronological age[62,103,397]. Puberty appears to be significant as far as attaining maximum bone mass in adulthood is concerned, since boys who do not receive adequate treatment sufficiently early show reduced bone density as adults and never make up this deficit[112].

In an atttempt to imitate the slow natural increase in testosterone, 50-100 mg testosterone enanthate is usually administered intramuscularly every 4 weeks. After a 6-month treatment period, a treatment-free interval of 2-3 months will clarify whether puberty has started spontaneously, as often happens in CDD, or whether endogenous testosterone is still dormant. Then, testosterone enanthate treatment should be restarted and increased in the second year of therapy to 100 mg every 3-4 weeks. In the third year the dose should be increased to 200-250 mg every 3-4 weeks. This procedure is well established and has been well documented[322,370]. Alternatively, if compliance is good, oral testosterone undecanoate may be given at 40 mg daily in year 2 and 80 mg daily in year 3[68]. The administration of other androgens or testosterone formulations in delayed puberty has not been documented in studies. It remains to be seen whether these will include applications for testosterone undecanoate or testosterone gel. If delayed puberty is hypophyseal in origin, human chorionic gonadotropin treatment, which affects LH, can be given. If the cause is hypothalamic, pulsatile GnRH could possibly be administered. However, both these treatments are more expensive than testosterone injections, which is why they are less frequently used.

The usual checks are performed during treatment. In addition, clinical signs of virilization such as secondary hair growth and growth of genitalia as well as special side-effects such as acne should be documented.

4.2.3. Excessive growth

Approximately 3 % of all boys are taller than 192.5 cm (= 6 feet 4 inches), and are therefore excessively tall. Usually, medical advice is not sought until the boy is over 2 m in height. The parents are often tall, too. Prediction of height is based on the methods of Greulich and Pyle or Tanner-Whitehouse, using detailed radiological analysis of wrist bones compared with a standard. Since this requires much expertise, the calculation of potential final height should be left to experienced clinics.

If the patient is suffering from psychological problems, this is an indication that excessive growth in height should be halted. There is rarely a medical reason for this treatment. Usually, boys are treated with 500 mg testosterone enanthate intramuscularly 1-2 times a week. Besides closing the epiphyses, testosterone administration initially increases growth rate, and therefore treatment should start before a bone age of 14 years if a significant reduction in height is to be achieved. If response to treatment is good, final height can be reduced by about 5-8 cm, rarely more. However, if treatment is started too late, final height can be inadvertently increased.

Occasionally, high-dose testosterone therapy leads to mild pretibial edema and weight gain due to water retention. More frequently, the patient develops acne and is forced to stop treatment. In very rare cases, patients develop acne fulminans, which requires dermatological treatment even after the-rapy has ended. Hypophyseal function is suppressed by exogenous testosterone, but this recovers fully when therapy is finished. Accordingly, final testis volume is normal and there is no evidence of endogenous testosterone production or fertility being impaired later on. No abnormal behavior such as increased aggression or excessive sexual activity has been observed[56,101].

4.2.4. Transsexuality (female-to-male)

Transsexuality is a condition in which patients with normal genetic and phenotypical gender differentiation are convinced that they belong to the opposite sex. They usually have an urgent desire for their body to match their psychological gender.

It is usually recommended that patients live first in the role of the opposite sex for 1-2 years before starting hormone therapy, as some of the changes are irreversible. Then follows a 1-2 year phase of hormone therapy before sex-change surgery is performed (mastectomy, removal of ovaries, hysterectomy and phalloplasty). This takes place in highly specialized clinics.

In female-to-male transsexuals hormone therapy is based on the application of testosterone. For this indication most experience exists in the use of either intramuscular application of testosterone enanthate or oral administration of testosterone undecanoate. Recently, the first clinical experiences with the use of intramuscular testosterone undecanoate (Nebido®) in female-to-male transsexuals were reported and showed very favourable results. The application of testosterone causes suppression of gonadotropins and ovulation stops. Menstruation also stops. Alternatively, oral testosterone undecanoate can be given at a dose of 160-240 mg per day. This is considerably less effective in terms of virilization and takes longer to work than testosterone enanthate. There is no experience in the use of other preparations. After the ovaries have been removed, testosterone substitution is continued for the rest of a patient's life.

Testosterone therapy causes virilization and increased growth of secondary hair. Over the course of several years this will resemble that of a normal male, and the patient will need to shave. Approximately a quarter of patients develop acne, which occasionally requires topical treatment. The voice becomes significantly and irreversibly deeper within 2-3 months. A decrease in subcutaneous adipose tissue together with an increase in musculature causes masculinization of body shape. Later on, there is an increase in visceral adipose tissue. The clitoris grows considerably in size and may be mistaken for a small penis. It is important to assess bone density regularly since, according to recent findings, testosterone therapy is not able to maintain bone mineral content in all patients[402].

4.2.5. Late-onset hypogonadism

Late-onset hypogonadism is a proven indication for androgen replacement therapy. To provide an adequate and coherent representation of this indication for testosterone substitution, an entire chapter (☞ Chapter 5.) has been devoted to this subject.

4.2.6. Primary and secondary osteoporosis

The prevalence of osteoporosis in men has significantly risen in recent years, and has enormous repercussions on health economics. After excluding secondary causes (hypogonadism, osteomalacia, primary hyperparathyroidism, alcohol abuse), many men are diagnosed with idiopathic or primary osteoporosis. There is no doubt that men with primary or secondary hypogonadism benefit from testosterone substitution[30]. The osteoanabolic effect of androgens in postmenopausal women is well known, although this therapy has been largely abandoned in favor of other medications because of the virilization that occurs in approximately one-half of female patients. Nevertheless, only androgens that are aromatizable produce an osteoanabolic effect. In men, too, estrogens which are formed in the peripheral tissues (e.g. adipose tissue) and bones from testosterone and other androgens are the critical stimulator of bone composition[317]. There is only one study that deals with the question of whether additional administration of androgens to men with osteoporosis, but without testosterone deficiency, has a favorable effect on bone density. In this, men with at least one fracture of the vertebral body due to idiopathic osteoporosis received an intramuscular injection of 250 mg testosterone ester every 14 days. After 6 months, bone density had increased by an average of 5 %. Parameters of bone density such as deoxypyridinoline, which were raised initially, normalized during therapy[13]. It is not clear whether this prevents fractures in the long-term. The positive correlation of serum estradiol with increase in bone density emphasizes the necessity of administering an androgen which can also be aromatized to estrogens. Because of the lack of data, the administration of androgens to eugonadal men with osteoporosis cannot be generally recommended, since there is clear evidence of a beneficial effect on fracture rates from other antiosteoporotics (e.g. bisphosphonates). However, if there are no contraindications to the use of these preparations (e.g. renal failure, intolerance), androgens provide a therapeutic alternative, especially since the anabolic 19-nortestosterone decanoate has a proven osteoanabolic effect in older studies[3,98].

In contrast, testosterone is indicated in the most common type of osteoporosis, which is induced by corticosteroids. Osteoporosis is one of the most common and most serious side-effects of corticosteroids. In addition to a direct osteocatabolic effect, high-dose corticosteroids induce secondary hypogonadism through negative feedback to LH secretion. This causes androgen deficiency, which exacerbates bone destruction. In a controlled study of asthmatics on long-term corticosteroid medication, administration of 250 mg testosterone ester every 4 weeks increased bone density by 5 %[313].

A recent well designed randomized, placebo-controlled study confirmed these findings[86]. Men on long-term glucocortocoid treatment received either intramuscular testosterone, nandrolone (which is poorly aromatizable) or placebo. Both androgens, testosterone and nandrolone, significantly increased muscle mass (+3.5% and + 5.8%, respetively) whereas in the placebo group muscle mass decreased (-0.9%). Lumbar spine bone mineral density increased only in men treated with testosterone (+ 4.7%)[86]. This study underlines the importance of concomitant therapy to prevent the deleterious side effects of glucocorticoids and the importance of aromatizability to achieve a beneficial effect on bone metabolism.

Therefore, men on long-term corticosteroid therapy should have their serum testosterone levels checked regularly. If these are too low, one should consider administering testosterone, e.g. 250 mg testosterone enanthate every 3 weeks. Administration of the anabolic nandrolone decanoate produced a 5 % increase in bone density in women with rheumatoid arthritis, whereas in the control group bone density fell by almost 7 %[3]. Therefore, women with corticosteroid-induced osteoporosis can also be started on androgens if there are contraindications to bisphosphonates.

4.2.7. Weight loss in consumptive diseases

Restricted physical capacity and diminished quality of life are characteristic of numerous consumptive diseases. Whereas the administration of androgens in malignoma has not become established, HIV infection is a new indication. In a randomized, placebo-controlled study, the intramuscular administration of 300 mg testosterone enanthate every 3 weeks caused a reversal in weight loss and an increase in body weight, lean body mass and muscle mass. This manifested as an increase in physical capacity and improved quality of life[137]. As the prognosis for HIV sufferers has considerably improved, testosterone treatment should be attempted in all patients with AIDS-associated weight loss. Moreover, testosterone therapy may have a beneficial effect on regional fat distribution, which is often disturbed in HIV sufferers. Anabolics (e.g. 19-nortestosterone, metenolone, clostebol) may be even more effective than testosterone, but this has not been verified by studies.

4.2.8. Hereditary angioedema

Hereditary angioedema is caused by a genetic deficiency in the C1 esterase inhibitor. It causes recurrent edema of the skin, gastrointestinal tract and respiratory tract. Alkylated androgens such as danazol, stanozolol or metanolone enanthate which stimulate hepatic production of the C1 esterase inhibitor can be administered prophylactically, but not during an attack. Testosterone and its esters do not display this effect. Danazol is administered at a dose of 400-600 mg per day and then reduced to the lowest effective dose[10,352].

4.2.9. Male contraception

Administration of high-dose testosterone suppresses hypophyseal production of the gonadotropins FSH and LH through negative feedback. As a result, the testes are not stimulated, and testicular endocrine and exocrine function ceases. Whereas the deficiency of endogenous testosterone is compensated by the administration of exogenous testosterone, suppression of FSH manifests as reduced or complete absence of spermatogenesis, resulting in severe oligospermia or azoospermia.

In the most extensive study of this subject carried out to date, intramuscular administration of 200 mg testosterone enanthate per week caused azoospermia in two-thirds of volunteers of Caucasian origin within 6 months. When therapy was continued, only one pregnancy occurred among 157 couples within a year, when no other form of contraception was used. Although this essentially underlines the efficacy of this procedure, weekly intramuscular injections are not practical for broad application, and persistent oligospermia in a third of volunteers is not tolerable for a proven application[430]. The onset of azoospermia may be accelerated and intensified by simultaneous administration of a GnRH antagonist (cetrorelix, Nal-Glu-GnRH) or a gestagen (levonorgestrel, cyproterone acetate, medroxyprogesterone acetate). Therefore, it is anticipated that efficacy rates will be higher in the future[431].

It is still not known whether high-dose testosterone therapy will have any adverse effect on the prostate in the long term. New synthetic androgens with very specific tissue and selective androgen effects (e.g. MENT) might find an application in this area[14]. At present, hormone contraception in men is regarded purely as an experimental procedure, especially since the medicolegal aspects concerning liability in treatment failures and side-effects have not been clarified.

4.2.10. Aplastic and renal anemia

Since recombinant erythropoietin became available, stimulation of hematopoiesis by androgens has largely been ignored. Androgen treatment can be attempted if erythropoietin therapy is unsuccessful.

Testosterone stimulates renal and extrarenal formation of erythropoietin and directly promotes differentiation of pluripotent stem cells to erythropoietic precursor cells. Numerous case reports show an improvement in hematopoiesis in individual cases of aplastic anemia, Fanconi's anemia, hemolytic anemias, myelofibrosis and sickle-cell anemia, although the success rate is very variable. In hemochromatosis with androgen deficiency, testosterone substitution leads to increased hematopoiesis. Consequently, the patient can be bled more frequently so the iron overload can be more rapidly corrected[10,117].

Testosterone esters such as testosterone enanthate or undecanoate may be administered. The usual substitution dose is administered initially and then the dose is slowly increased.

4.2.11. Abnormally small penis (micropenis) in the newborn

The term micropenis refers to a penis of normal structure and shape but which is too small. The penis of a healthy newborn male is 3.9 ± 0.8 cm when extended. A penis which is less than 2 cm in length is considered to be a micropenis. This condition is often caused by primary or secondary hypogonadism, chromosome defects or rare syndromes.

The success of testosterone or dihydrotestosterone therapy in boys with abnormally small penises is disputed. As the condition is very rare, there are no studies to support this premise. It is well known that the penises of adults contain significantly fewer androgen receptors than those of young or adolescent boys. Therefore, if the penis needs to be lengthened, androgen therapy should be started before the onset of puberty. A dose of 25 mg testosterone enanthate intramuscularly every 2 weeks can cause significant growth in the penis of a newborn. However, there is partial regression after treatment has ended. Topical application of testosterone or dihydrotestosterone in the form of gel is no more effective. It has not been proved whether such therapy actually has a beneficial effect on the overall length of the penis in adults. If parents insist on therapy, testosterone treatment can be given, particularly since no adverse effect has been described[2,255].

4.2.12. Non-indicated applications of androgens

Testosterone plays an important role in stimulating and maintaining spermatogenesis. Therefore, androgen treatment was one of the first treatments to be used in men with idiopathic infertility. However, recent studies have shown that neither mesterolone nor high-dose testosterone undecanoate, nor testosterone rebound therapy, has a beneficial effect on fertility in men suffering from idiopathic infertility[76,214,432]. As a result, these treatments should not be used, as they inadvertently delay the use of more effective procedures (e.g. assisted fertilization), raise false hope and are costly.

The administration of testosterone or other androgens to men with erectile dysfunction and normal testicular endocrine function is very rarely successful, and is mainly based on a placebo effect[19]. If sufficient endogenous testosterone is produced, additional administration of exogenous testosterone will not act as an aphrodisiac nor will it increase potency (otherwise testosterone would be one of the most commonly administered preparations). In addition, one should bear in mind that exogenous testosterone suppresses endogenous testosterone production because of negative feedback to the hypophysis, and therefore there is no additive effect.

However, recently several trials have shown beneficial effects of testosterone substitution in men with erectile dysfunction, who do not respond to phosphodiesterase-5-inhibitors (PDE-5-I)[351]. In a randomized trial sildenafil non-responders receiving transdermal testosterone showed a significant increase in penile arterial blood flow and simultaneously an improvement of erectile function domain score at IIEF (International Index of Erectile Function) over placebo[17,350]. This has been confirmed in men with diabetes mellitus type 2, who did not respond to sildenafil. Pre-treatment with oral testosterone undecanoate significantly improved response to sildenafil[55,204]. Thus, in case of several unsuccessful attempts of intercourse despite use of PDE-5-I in the highest applicable dose, a trial of testosterone substitution in addition to the PDE-5-I might be helpful.

If testosterone deficiency is excluded in men with erectile dysfunction, there should be further clarification of the cause and, if possible, causal therapy should be given. If applicable, phosphodiesterase inhibitors are now also available. These are particularly effective in psychogenic potency disorders.

4.3. Evaluation procedures before and during treatment with androgens

Before starting androgen administration, the diagnosis of each patient should be confirmed. Even a single dose of testosterone can affect subsequent hormone measurements, so that it might not be possible to make a diagnosis for some time. Therefore, one should always await a diagnosis before starting androgen administration, especially since a delay of a few days is unlikely to be harmful.

To exclude contraindications and ascertain baseline values, the following should always be available before starting therapy: blood count, liver transaminases (GOT and GPT), bilirubin and, in men over 50 years of age, prostate-specific antigen (PSA). Above the age of 50, palpation and transrectal ultrasound (TRUS) of the prostate are included in the pre-therapeutic diagnosis, since focal changes should be excluded/eliminated before androgen administration.

Osteodensitometry should be carried out to determine bone mineralization in patients with primary or secondary hypogonadism and older men. During substitution therapy, bone density should be repeated after a minimum of 1 year, to monitor the success of therapy. It is pointless to take more frequent measurements, because changes in bone mineralization are slight and osteodensitometer measurements can be inaccurate.

Before therapy	After 3 months	After 6 months	After 12 months
• Case history • Examination • Testosterone • LH • Blood count • HDL, LDL • Liver values • Bone density	• case history • examination • testosterone • LH • blood count • HDL, LDL • liver values	• case history • examination • testosterone • LH • blood count • HDL, LDL • liver values	• case history* • examination* • testosterone* • LH* • blood count* • HDL, LDL* • liver values* • bone density†
In men over 50 years of age additionally			
• PSA • TRUS or palpation	• PSA	• PSA • TRUS or palpation	• PSA* • TRUS or palpation†

Table 4.6: Diagnostic parameters before and during androgen therapy, then 6-monthly* and yearly†. LH = luteinizing hormone; HDL = high-density lipoprotein; LDL = low-density lipoprotein; PSA = prostate-specific antigen; TRUS = transrectal ultrasound of the prostate.

The first follow-up examination should take place after 3 months (☞ Table 4.6). The patient will be asked about tolerance and effect with regard to physical capacity, libido, potency and general well-being. If a questionnaire was used to assess the symptoms prior to testosterone substitution, it can be useful in the follow-up for evaluating the efficacy of therapy. Clinical effect is of primary importance to dose adjustment. Physical examination will concentrate particularly on androgen target organs. Changes in secondary hair growth, musculature, body shape, genitals and voice will be documented. A blood count will monitor the success of therapy in androgen-deficiency anemia and exclude polyglobulism[264].

Trough blood samples should be taken, i.e. with intramuscular injections, at the end of an injection interval immediately before the next dose when serum testosterone should be at the lower limit of normal. In patients with primary hypogonadism, serum LH should additionally be determined to evaluate substitution. Ideally, LH which was increased before therapy should have normalized on testosterone substitution. However, this practically never occurs with oral testosterone or testosterone patches, and rarely happens with intramuscular testosterone enanthate. In the future, leptin may be a suitable indicator of the quality of testosterone substitution. After examining the testosterone trough level, serum LH, blood count and clinical effect simultaneously, the dose of testosterone or administration interval can be adjusted.

To improve compliance, it may be useful to supply the patient with a booklet in which the recorded findings are entered. These can be updated at each check-up.

4.4. Contraindications to androgen substitution

Confirmed prostate carcinoma is an absolute contraindication for androgen therapy (☞ Table 4.7). If the pathological prostate findings are unclear, one should await a definitive diagnosis before starting androgen treatment. In addition to PSA testing, transrectal sonography is recommended in men over 50 years of age to determine size and evaluate any focal changes in the prostate. It is not known whether androgens may be administered several years after removal of a T1 carcinoma if there have been no recurrences. The indication would have to be very carefully weighed up against the potential risk.

Mammary carcinoma in men is also a contraindication. Androgens are aromatized to estrogens and therefore the growth of estrogen receptor-positive mammary carcinomas is promoted by androgen therapy. On the other hand, hypogonadism-induced gynecomastia is not a contraindication, and is more likely to regress on testosterone substitution.

Absolute contraindications	Relative contraindications
• Prostate carcinoma • Unclear prostate findings • Mammary carcinoma • Polyglobulism	• benign hypertrophy of the prostate

Table 4.7: Absolute and relative contraindications to androgen therapy.

Benign hypertrophy of the prostate is a relative contraindication. If there is already significant hypertrophy with the formation of residual urine in the bladder, caution should be exercised. If the indication is urgent (e.g. severe osteoporosis in androgen deficiency), low doses can be administered initially and the prostate monitored by sonography, and PSA, residual urine and urinary flow measured.

If there is polyglobulism with increased hematocrit and a high red cell count, a general examination should be carried out first and appropriate treatment administered before androgens can be initiated. Androgens stimulate erythropoiesis and increase any polyglobulism present, resulting in poor blood flow (rheology) and a risk of thrombosis. If polyglobulism develops during androgen treatment, treatment should be discontinued until the red cell count, hematocrit and serum hemoglobin have normalized and then androgen treatment can be restarted at a lower dose.

4.5. Side-effects

Testosterone substitution and treatment are very safe, and associated with few side-effects (☞ Table 4.8). This is especially true of testosterone and its esters[125,126]. On the other hand, anabolics and 17α-alkylated androgens produce significantly more side-effects and can be especially hepatotoxic. As androgens have a sebum-stimulating effect, acne frequently develops during androgen therapy, especially at the start of treatment and after a long period of androgen deficiency. In the majority of cases this is transient, does not require treatment and ceases spontaneously. In extremely rare cases on very high-dose testosterone therapy, or in anabolic abuse, the special form of acne called acne fulminans develops in patients with a familial predisposition, forcing treatment to be withdrawn.

Acne	rarely at start of therapy
Gynecomastia	rarely
Priapism	extremely rarely
Sleep apnea syndrome	in men with a predisposition
Polyglobulism	in men with a predisposition
Hepatotoxicity	only 17α-alkylated androgens

Table 4.8: Potential side-effects of androgen therapy. HDL = high-density lipoprotein; LDL = low-density lipoprotein.

17α-Alkylated androgens can induce dose-dependent peliosis hepatis (hemorrhagic liver cysts) or jaundice, and benign and malignant liver tumors which is why these preparations are obsolete in Europe[60]. Testosterone and testosterone esters do not result in these side-effects, although on extremely rare occasions liver transaminases can be raised, especially in men with pre-existing liver disease. Discontinuing therapy for a while may clarify the link with testosterone.

Gynecomastia may develop during testosterone therapy. This presumably results from a temporary imbalance between androgens and estrogens as in puberty. As therapy continues the gynecomastia usually recedes.

Priapism occurs so rarely during testosterone therapy that a causal connection is not suspected. Over a 20-year reporting period, the WHO Collaboration Center for International Drug Monitoring in Uppsala, Sweden, has not seen a connection between priapism and testosterone therapy, and only isolated cases are described in the literature.

In very rare cases, testosterone therapy can induce sleep apnea syndrome in predisposed men (e.g. those with pronounced obesity and chronic obstructive pulmonary disease), but this is reversible. Affected patients display an excessive increase in hematocrit (over 50 %) and polyglobulism. When testosterone therapy is discontinued, sleep apnea syndrome and polyglobulism regress[249,328]. Only one case of sleep apnea syndrome has been reported in studies of the efficacy of androgen therapy in older men. Nevertheless, caution should be exercised when administering androgen therapy to patients with predisposing factors, and they should be questioned about their sleep patterns while undergoing substitution therapy.

Androgens may reduce HDL cholesterol and increase LDL cholesterol independently of the dose and route of administration. Whether this change in atherosclerotic risk actually affects morbidity and mortality is unclear. However, life expectancy in men is significantly less than in women, and the incidence of cardiovascular disease in men is higher than in women. Therefore, hypogonadal men on testosterone substitution should have their lipid metabolism monitored. If LDL cholesterol is raised, they should be started on treatment for hypercholesterolemia.

It is said that androgens increase aggression. However, neither studies in patients on testosterone substitution for hypogonadism nor studies in men on high-dose testosterone therapy for contraception reveal an increase in aggressive behavior due to testosterone. The administration of testosterone, however, does make men more active and improves pre-existing depression and frustration[393,416]. Libido, interest in sex and activity increase to levels similar to those in healthy men. Therefore, it is anticipated that androgens will have a beneficial effect on behavior when administered for hypogonadism.

4.6. Abuse of androgenic anabolic steroids

Abuse of anabolics is extraordinarily widespread in competitive sport, as well as in certain leisure sports (sports involving strength and body-building). The trend was started by weight-lifters in the 1950s, and has continued to increase since then. It is estimated that between 2 and 7 % of international athletes and between 7 and 10 % of college students and Americans practicing leisure sport take anabolics. No data are available for Europe, but according to the experiences of the author in 'Sport and Fitness studios', the use of anabolics is common.

Whereas the international literature on sports medicine has always refuted the use of anabolics, based on their own experiences and the observations of top sportsmen (e.g. the 100-m runner, Ben Johnson), users of anabolic steroids have been convinced for some time of the performance-enhancing effect of anabolics. This was confirmed by a doping program in the former East Germany which was only discovered after reunification, as well as a recently published study[41]. In the body-building 'scene' there is specialist literature describing the effects, side-effects and application protocols in detail. Previously, usually one preparation was taken continuously or intermittently during training. Today, combinations of different oral and intramuscular anabolics at 10-100 times the therapeutic dose are gradually established over 6-12 weeks, then kept at the highest dose for 4-8 weeks then reduced again ('stacking'). The most popular preparations have remained unchanged for years, presumably due to their proven efficacy: nandrolone, methandienone, stanozolol, methenolone and methyltestosterone (☞ Figures 4.5 and 4.6). Anabolic steroids are often combined with other medications such as thyroid hormones (reduction of adipose tissue), antiestrogens (prevention of gynecomastia), diuretics (reduction of water retention and dilution of urine to reduce the concentration of steroid in the event of a drug test) and anabolic growth hormone. The use of anabolics which are not medically indicated is to be condemned, not only for ethical reasons, but because serious side-effects can occur making intake dangerous. In addition, because of their prescription-only status, anabolics are largely purchased on the black market or through the internet from unknown chemical laboratories, and therefore the quality and purity of the product are uncertain. Veterinary medical products are also used. Anabolics suppress gonadotropins and therefore cause atrophy of the testes, which decrease in size significantly (☞ Table 4.9). Spermatogenesis stops, resulting in reversible oligospermia or azoospermia. After discontinuing anabolics, spermatogenesis can take several months to recover, depending on the duration of administration. Premature epiphyseal closure in adolescents may cause final height to be reduced. Aromatizable anabolics almost always cause gynecomastia because of the considerable increase in estrogens. After discontinuing anabolics, gynecomastia often does not completely regress, which is why patients occasionally demand a mastectomy. Hepatotoxic side-effects occur after intake of 17α-alkylated ana-

Male reproductive system	Female reproductive system
• Atrophy of testes • Oligospermia, azoospermia • Infertility • Gynecomastia • Testosterone deficiency	• Masculine secondary hair growth • Oligomenorrhea, amenorrhea • Infertility • Virilization, male appearance • pitch of voice drops • Hypertrophy of the clitoris
Both sexes	
Hematology	Cardiovascular
• Erythrocytosis • Polyglobulism • Risk of thrombosis	• Myocardial hypertrophy • Disturbed contractility • Increased blood pressure
Liver	Metabolism
• Cholestatic jaundice • Peliosis hepatis • Hepatocellular adenoma and carcinoma	• Reduction of HDL cholesterol • Increase in LDL cholesterol • Hyperinsulinemia
Skin	Musculoskeletal
• Acne • Androgenic alopecia	• premature epiphyseal closing with reduced final height
Psychological	
• Euphoric effect with risk of developing dependence • Depression when drug discontinued • Mood swings	

Table 4.9: Side-effects of anabolic abuse.
HDL = high-density lipoprotein; LDL = low-density lipoprotein.

bolics (e.g. stanozolol, methyltestosterone, oxandrolone) and manifest as jaundice, peliosis hepatis and liver tumors. In addition, anabolic abuse causes a considerable reduction in HDL cholesterol and an increase in LDL cholesterol. It is not known whether the long-term intake of anabolics has consequences for cardiovascular disease. Finally, anabolics also promote hypertrophy of cardiac muscle cells, resulting in an enlarged heart and disturbed contractility. This might explain the isolated case reports of cardiomyopathies and sudden heart death in anabolic abuse. Isolated cases of severe depression with suicide and psychoses have been described following abrupt discontinuation of anabolics. In women, anabolics additionally cause irreversible virilization, hypertrophy of the clitoris and lowering of the voice[435].

In competitive sport, steroid abuse is detected by analyzing a steroid profile using gas chromatography/mass spectrometry. This provides definite proof of abuse. Testosterone intake is determined by measuring the ratio of testosterone to epitestosterone in the urine. Normally, the ratio of the steroids is 6 : 1 or less, and increases sharply with testosterone intake. In clinical practice, a severely reduced HDL cholesterol level combined with high-normal or increased hematocrit and suppressed gonadotropins in men with very athletic build is indicative of anabolic abuse. The suspicion is strengthened by low testosterone levels, small testes, oligospermia or azoospermia and acne. Before further diagnosis or testosterone therapy, details of anabolic intake should be discussed with the patient.

Androgen deficiency in older men

5. Androgen deficiency in older men

5.1. Age-associated changes in androgen production

Testosterone production decreases as men become older. Numerous cross-sectional studies show an inverse relationship between age and serum testosterone level (☞ Figure 5.1).

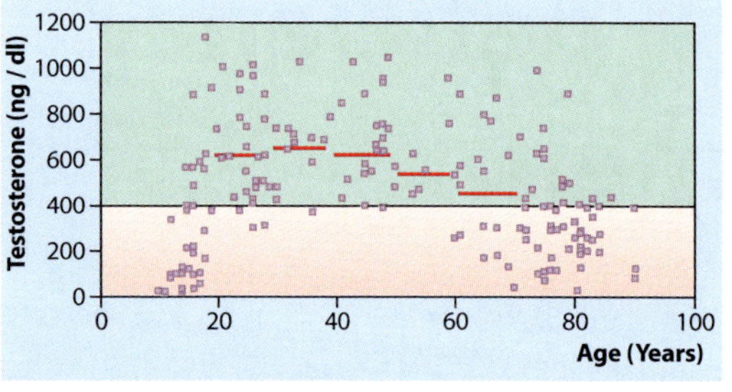

Figure 5.1: Cross-sectional study of serum testosterone levels as a factor of age in healthy men. Between the ages of 20 and 40 testosterone levels reach a plateau in healthy men, then they start to fall. Obviously in a high proportion of older men, serum testosterone levels are below the normal range for young men. Nevertheless, many older men have completely normal serum testosterone levels. The horizontal lines give the mean for each decade[410].

Between the ages of 40 and 70 years, biologically active free testosterone in the serum falls by approximately 1.2 % per year. There is a similar simultaneous increase in SHBG (☞ Figure 5.2).

In the Baltimore Longitudinal Study of Aging (BLSA), 20 % of men over 60 years of age, 30 % of men over 70 years of age and 50 % of men over 80 years of age had a low total testosterone level. When free, biologically active testosterone was taken into account, the proportion of hypogonadal men was even higher (☞ Figure 5.3)[40,109,130,131,148,384,411].

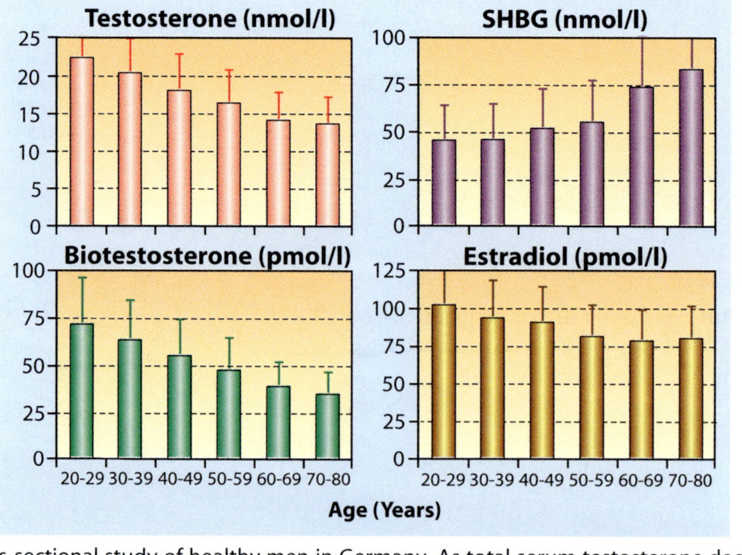

Figure 5.2: Cross-sectional study of healthy men in Germany. As total serum testosterone decreases, sex hormone-binding globulin (SHBG) increases to the same extent. Therefore there is a sharper reduction of biologically active testosterone than of total testosterone. Serum estradiol falls in parallel with the decrease in androgens since it is formed by aromatization from the prohormone testosterone[232].

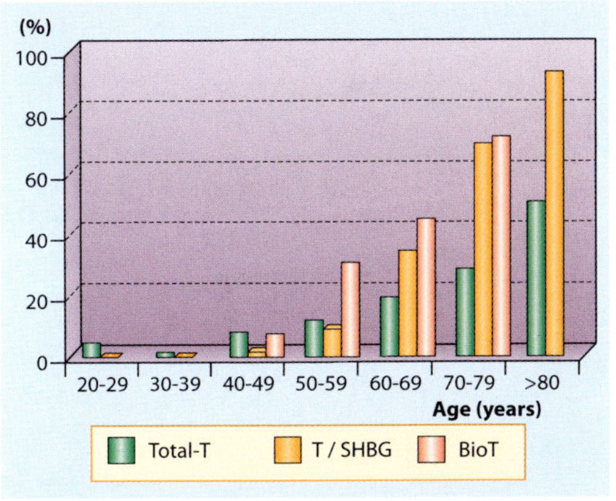

Figure 5.3: Percentage of men with low testosterone, testosterone/sex hormone-binding globulin (SHBG) ratios or bioavailable testosterone. The increase in SHBG makes it easier to identify men with testosterone deficiency. Data according to reference 386.

Serum estradiol levels may fall in older men as a result of androgen deficiency. This is not an independent phenomenon but a consequence of the testosterone deficit. Testosterone is the prohormone for estradiol. The enzyme aromatase is present in a number of tissues, especially adipose tissue, and this can metabolize testosterone to estrogens (Section 1.3.3.).

Many of the changes associated with age such as loss of bone and muscle mass, loss of body hair, low hemoglobin, increased fat mass and deterioration in physical, sexual and cognitive capacities are very similar to symptoms of androgen deficiency in classical forms of hypogonadism in younger men. It is therefore suspected that there is a close connection between androgen deficiency in older men and symptoms of old age. However, no causal connection has yet been proven. Independent of whether androgen deficiency is actually responsible for these symptoms in older men, the effect of androgen substitution in older men is being intensively researched. One aim of therapy is to correct clinical symptoms such as osteoporosis, muscle atrophy and low hemoglobin concentrations. Another is to improve quality of life and subjective symptoms such as hyperhidrosis, loss of libido, lack of drive, and decrease in mental and cognitive powers associated with growing older. However, bearing in mind demographic developments and the continuing increase in the numbers of old and elderly men throughout the world, research is also aimed at maintaining independence and preventing the need for care and nursing. Serum testosterone concentrations correlate positively with muscle mass, but also with the capacity to perform everyday tasks independently and the requirement for care[58].

There are various terms to describe androgen deficiency in older men. The terms 'virile climacterium' and 'andropause' which falsely suggest an analogy to the menopause in women, are no longer used (☞ Table 5.1).

Criterion	Menopause	Androgen deficit
Onset	abrupt within 1-2 years	gradual, over decades
Extent	complete loss of estrogen	partial androgen deficiency
Frequency	all women	approximately 20–35 % of men
Cause	primary hypogonadism	combination of secondary and primary hypogonadism

Table 5.1: Dissimilarities between androgen deficit in older men and menopause in women.

'Andropenia' is purely a descriptive term. Intentionally the term 'partial androgen deficit in aging men' (PADAM) is not uncommon. However, recently the term 'late-onset hypogonadism' has been coined and probably describes the situation best. Although androgen deficiency has not been proved to be causally responsible for the complex of symptoms of reduced 'virility', the question is whether androgen substitution can improve clinical symptoms such as muscle atrophy, reduced functional capacity, osteopenia/osteoporosis, loss of libido and mood, and whether it has a favorable effect on quality of life, morbidity and mortality.

Available results from studies carried out to date comprise treatment phases of up to 3 years and show positive effects without appreciable side-effects, as described below. However, testosterone therapy in the treatment of late-onset hypogonadism is currently still experimental, and long-term therapy is not yet supported by data. Therefore, particular care is required in defining the indication, excluding contraindications and documenting symptoms[39,40,386].

Andropenia of old age appears to be caused by a hypothalamic dysregulation. Several studies with detailed analysis of pulsatile LH secretion indicate

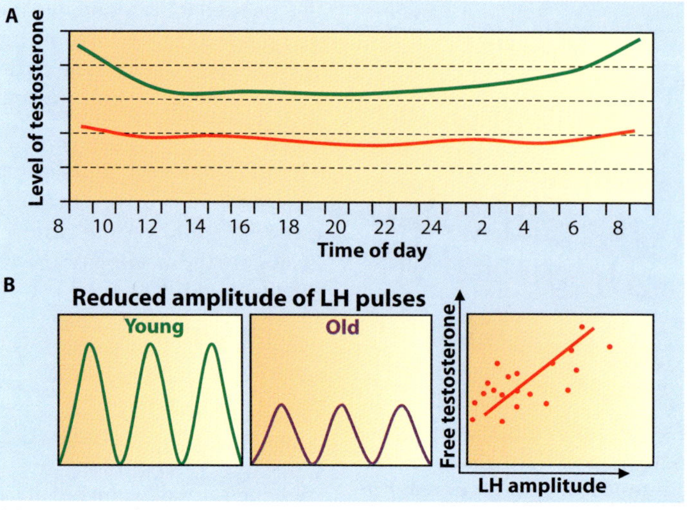

Figure 5.4: (a) In older men (red) the morning increase in serum testosterone characteristic of the circadian rhythm in young men (green) is absent. (b) This is caused by a reduced pulse amplitude with unchanged luteinizing hormone (LH) pulse frequency so that only half as much LH is secreted as in young men. Pulse amplitude correlates with free testosterone. Schematic representation according to data from references 57 and 406.

a reduction in GnRH secretion and support the concept of hypothalamic secretions becoming exhausted with increasing age[261,406,409], as is seen with other endocrine systems (e.g. somatopause, growth hormone deficiency). At the same pulse frequency, the pulse amplitude of LH is reduced, and therefore less LH is secreted (☞ Figure 5.4).

Moreover, testosterone secretion no longer follows a daily rhythm. Whereas in young men the highest serum testosterone levels are in the early morning hours and during the morning, falling away by evening, testosterone levels in older men remain at the low evening level and do not fluctuate much around this[57]. Reduced secretion of LH, which is formally equivalent to secondary hypogonadism, is attributed to reduced production of GnRH and increased sensitivity to the negative feedback of testosterone. In addition, reduced testosterone production is promoted by a reduced response by the Leydig cells to stimulation by LH[40,404,405]. This typically results in a clinical finding of reduced testosterone combined with normal LH levels.

It is not clear whether reduced testosterone production is the result of physiological aging processes or the expression of a pathology. General illnesses are more frequently the cause of reduced serum testosterone levels[130,270], so there is obviously a connection between general health and production of testosterone[194,195]. Therefore, reduced testicular and hypothalamic function may be indicative of a pathological finding, e.g. atherosclerotic vascular changes with hypoxia-induced damage.

5.2. Clinical symptoms in older men (testosterone deficiency syndrome)

The clinical symptoms of older men comprise objective somatic criteria and changes, subjective symptoms and feelings of ill-health which have no disease status. It is not clear whether there is always a causal connection with androgen deficiency since there is no single pathognomonic symptom for testosterone deficiency. Nevertheless, when the individual symptoms are combined, androgen deficiency can often be deduced (☞ Table 5.2).

Androgen target organ	Change	Clinical signs
Bones	osteopenia, osteoporosis	pain in bones, fractures
Musculature	atrophy	neurasthenic weakness, exhaustion
Adipose tissue	visceral adipose tissue ↑	weight increase, obesity, lipomastia, metabolic effects?
Erythropoiesis	anemia	chronic fatigue
Libido	loss of libido	sexual activity ↓
Potency	erectile dysfunction	impaired potency

Table 5.2: Somatic changes associated with age in man.

Many men above the age of 50 or 60 suffer from clinical signs of reduced 'virility'. Decreased muscle power, physical capacity, sexual function and osteopenia often lead to morbidity and reduced quality of life in older men, and this section of the population is growing all the time[39]. Psychological pressure is increased by an increased expectation to be active and physically capable, and therefore any restriction due to age is accepted with reluctance.

Clinical signs of andropenia in older men range from reduced muscle mass (sarcopenia) to muscle atrophy with reduced muscular strength and physical capacity. A 70-year-old man will have approximately 12 kg less lean body mass, and most of this is represented by muscle. Recent epidemiological studies show that the composition of the body starts to change after the age of 30. The non-muscle portion of lean body mass remains the same, but muscle mass decreases by 35-40 % between the ages of 20 and 80[27,61]. Muscle atrophy results from a decrease in the number and thickness of muscle fibers and a simultaneous reduction in the rate of protein synthesis, which eventually results in catabolism. In addition, physical capacity in terms of absolute strength and elasticity of muscle cells is less than in youth[353].

Because of this reduction in muscle mass and strength, older men suffer from restricted physical capacity and increased tiredness, and they become exhausted much more quickly than when they were younger. Moreover, muscle mass is an important determinant of bone density and fracture rate[326]. Muscle atrophy has an adverse effect on the ability of older men to care for themselves and they become dependent[61,226]. Sarcopenia is caused by a number of factors, not only a deficiency of testosterone. Physical activity and muscle training generally decrease with age. In addition, other anabolic hormones such as growth hormone and insulin-like growth factor (IGF-I) are absent.

Despite the sarcopenia, body weight usually increases owing to a simultaneous increase in fat mass of 18-36 %[320]. Adipose tissue increases, especially intra-abdominal visceral adipose tissue[39]. In men there is a link between low serum androgen concentrations and the amount of central abdominal fat ('android')[48,210,228,241]. No causal connection has been proved between the increase in visceral adipose tissue and metabolic changes such as increasing insulin resistance.

Androgen deficiency causes reduced bone mineralization, resulting in osteopenia or osteoporosis, which manifests as pain (backache) and myogelosis, reduced bone density and even osteoporotic fractures (☞ Table 5.3). An increasing number of men are diagnosed with osteopenia and osteoporosis. Approximately 25-30 % of all femoral neck fractures are in men. There are not always obvious risk factors for the development of osteoporosis, e.g. alcohol consumption, systemic diseases, medication or immobilization[169,318]. The three most common causes of osteoporosis in men are alcohol consumption, corticosteroid therapy and hypogonadism[46,293,318]. In older men in particular, reduced testosterone production is a crucial causal factor of osteopenia[12,170,291,326]. Two studies of male inhabitants of old people's homes detected a low serum testosterone level in almost 65 % of patients with femoral neck fractures, whereas a low serum testosterone level was detected in only 22 % of controls who had no fractures[170,371]. In several large epidemiological cross-sectional studies, a positive correlation was shown between free testosterone and its metabolite free estradiol, and bone density of the spine and femoral neck[40]. These findings mean that androgen deficiency is likely to be a causal factor in osteopenia/osteoporosis of older men.

Osteoporosis
• Backache
• Fractures
• Decrease in height
Body composition
• Increase in adipose tissue
• Lipomastia, gynecomastia
• Decrease in lean tissue
Muscle strength
• Muscle atrophy
• Diminishing strength
• Neurasthenic weakness
Skin
• Dryness
• Lack of sebum production
• Reduced secondary hair growth
Sexual function
• Loss of libido
• Erectile dysfunction
Anemia
• Chronic fatigue
• Neurasthenic weakness
Vegetative and subjective symptoms
• Hyperhidrosis
• Hot flushes
• Lethargy
• Lack of enthusiasm
• Apathy
• Lack of self-confidence
• Anxiety
• Depression
• Lack of perspective
• Sleep disorders

Table 5.3: Typical clinical symptoms in older men.

Erythropoiesis also decreases with age, so the number of red cells and the serum hemoglobin concentration fall to the lower third of the normal range or below[391]. Low hemoglobin concentration is associated with a reduced capacity to transport oxygen. This can lead to chronic fatigue, and intensify the deterioration in physical capacity resulting from muscle atrophy. It is not known whether the blood supply to other organs, e.g. heart and CNS, is impaired by this reduced ability to transport oxygen, and whether this also has an effect on function.

For many men, decreased sexual function and activity are their main subjective symptoms. Libido and impaired potency are frequent problems in older men. Whereas men in their 30s and 40s are sexually active approximately twice a week, this decreases to about twice a month in healthy men over 60[378]. According to the Massachusetts Male Aging Study, 35 % of all men over 40 years of age complain of impaired potency, which frequently causes great psychological pressure[267]. Approximately one-half of all men over 60 suffer from erectile dysfunction and approximately 15 % are completely impotent.

Nowadays, erectile disorders are usually considered to be caused by organic disorders, often a combination of several conditions. Conditions which are more common with age such as diabetes mellitus, atherosclerotic vascular changes, chronic renal failure, heart disease and hypertension are as damaging to potency as neurological, vascular and drug-induced disorders[193,194,220]. Accordingly, androgen deficiency is rarely the only cause of erectile dysfunction. Only approximately 10 % of all men with erectile dysfunction as their main symptom have low testosterone levels, and the incidence of low testosterone levels is no different in men with or without erectile dysfunction[218]. Therefore, testosterone therapy does not improve sexual function in men with normal testosterone levels[73,395]. On the other hand, some older men experience a significant improvement in sexual activity, libido and sexual satisfaction as the androgen deficiency is compensated[73,143,173]. However, success of therapy is very variable, and depends on the accompanying circumstances. Indeed, a meta-analysis of testosterone therapy in potency disorders concluded that the administration of testosterone to men with a low testosterone level is associated with a higher success rate than the administration of placebo; however, the studies on which these data are based have limited significance as most are poorly controlled and involve only small numbers of cases[173].

With the advent of phosphodiesterase-5-inhibitors (PDE-5-I) erectile dysfunction can be treated with a high success rate. In general, about 60 to 70% of the patients respond to PDE-5-1. However, the remaining one third of the patients do not respond to PDE-5-I. A randomized, placebo-controlled trial has recently shown, that non-responders to PDE-5-I may benefit from the concomitant therapy with testosterone. The application of transdermal testosterone has not only restored low or borderline serum testosterone to normal, but also increased penile arterial blood flow. This led in conjunction with sildenafil to a significant improvement of erectile function domain score at IIEF (International Index of Erectile Function) over placebo[17]. The observation, that PDE-5-I non-responders will definitively gain from the additional therapy with testosterone has been confirmed in several other studies[55,204,350].

Men with diabetes mellitus type 2 have especially benefited from the additional testosterone substitution[55,204]. Therefore, PDE-5-I non-responders should be given a chance to try additional testosterone substitution[351]. As in any situation of testosterone substitution, contraindications (e.g. prostatic disease) must be excluded prior to initiation of therapy.

It is not clear to what extent reduced cognitive abilities, poor concentration and general mood are connected with androgen levels. Brain cells express the androgen receptor and possess aromatase, so estradiol, which is a metabolite of testosterone, also plays a central role. In cross-sectional studies, a positive correlation was observed between free testosterone and improved cognitive function in older men[436]. A link was also determined between decreasing levels of free testosterone and cognitive abilities in the Baltimore Longitudinal Study of Aging (BLSA)[262]. However, it was not possible to make a definitive analysis of the effects of testosterone using established methods (neuropsychological test inventories). Therefore, it is difficult at present to attribute subjective symptoms to androgen deficiency and to deduce a therapeutic effect. The same applies to the effect of testosterone on aspects of mood (anger, grief, aggression). Androgens improve positive moods and reduce negative moods in young men with classical forms of hypogonadism[7,8]. They may also improve depression[416]. In smaller studies, testosterone substitution was found to have a positive effect on patients' own assessment of mood including sadness, energy, anxiety, tiredness, cheerfulness and nervousness. Nevertheless, there is a lack of convincing evidence to confirm the clinical impression that testosterone improves mood in older men, too[143,385]. In addition, interactions of different symptoms such as depression and erectile dysfunction must be taken into account, particularly since testosterone deficiency is frequently found with depression[348].

Androgen deficiency manifests in reduced sebum production, leading to dry, sensitive skin. Paradoxically, older men with testosterone deficiency often complain of hot flushes and hyperhidrosis. Although presumably the expression of a vegetative symptom, the cause of this has not been clarified. Secondary hair growth also decreases, and the patient complains of reduced body hair.

5.3. Clinical symptoms and diagnosis of androgen deficiency in older men

The diagnosis comprises a case history, including questions about the patient's sex-life, and professional and private lives, as well as a detailed general physical examination to evaluate the androgen target organs.

It is easier to lead into a discussion on case history if a symptom questionnaire is used. Two questionnaires have been developed for the purpose: the St Louis University Androgen Deficiency in Aging Males (ADAM) questionnaire (☞ Table 5.4) and The Aging Males' Symptoms rating scale questionnaire, which has been validated and translated into 14 languages (☞ Table 5.5) [149,272].

A high symptom score in either questionnaire should lead to hormone analysis. In addition, questionnaires make it easier to keep a systematic record of symptoms and clarify the main symptoms affecting the patient, and are suitable for recording any changes during therapy. The case history should cover all aspects of health, i.e. neurological (e.g. Parkinson's disease) and general diseases (e.g. cardiovascular risk factors, kidney and liver disease, etc.), so that a comprehensive picture of the patient's health can be ascertained. The same applies to the physical examination which should include auscultation, percussion, blood pressure and pulse (☞ Figure 5.5).

1. Do you have a decreased libido (sex drive)?
2. Do you lack energy?
3. Do you have less strength and/or endurance?
4. Have you lost height?
5. Have you noticed a decreased 'enjoyment of life'?
6. Are you sad and/or irritable?
7. Are your erections less strong?
8. Have you noted a recent deterioration in your ability to play sports?
9. Do you fall asleep after dinner?
10. Has there been a recent deterioration in your work performance?

Table 5.4: The St Louis University Androgen Deficiency in Aging Males (ADAM) questionnaire. All questions are to be answered yes or no. If the response to questions 1 or 7 plus three further questions is yes, the patient has androgen deficiency (88 % sensitivity, less than 60 % specificity)[272].

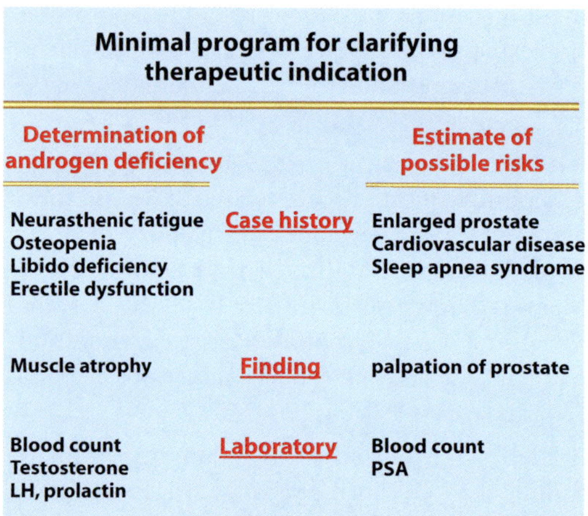

Figure 5.5: Minimum diagnostic program according to international recommendations for ascertaining androgen deficiency and excluding contraindications to testosterone substitution[265,386].

Evidence of androgen deficiency should be investigated by determining SHBG, LH and morning testosterone. It is not normally necessary to determine free testosterone, DHT or estradiol (☞ Figure 5.5). Indeed, it would be much more sensible to measure the biologically active part of total testosterone (free testosterone), especially since this is more sensitive in identifying men with androgen deficiency; however, assays used in clinical laboratories to determine free testosterone are not sufficiently accurate. Therefore, guidelines advise against using the determination of free testosterone in clinical testing[265,386]. If one wants to calculate free testosterone, this can be done from the values for total testosterone, SHBG and total protein (or albumin), which give an index of free testosterone equivalent to the true value of free testosterone (determined by equilibrium dialysis)[278,427]. The International Society for the Study of the Aging Male has a calculator on its internet page (http://www.issam.ch/ freetesto.htm). If total testosterone falls below 200 ng/dl (7 nmol/l), the diagnosis of hypogonadism is confirmed. If LH is low or low-normal, prolactin should also be determined to exclude prolactinoma. If prolactin is increased (> 60 ng/ml), an MRI scan should be taken of the sellar region. If total testosterone is between 200 and 400 ng/dl (7-12 nmol/l), the free androgen index can be calculated, or free testosterone using the dialysis method or bioavailable testosterone using ammonium sulfate precipitation[386]. The upper limit of substitution is considered to be 12 nmol/l. Above this level, hormone analysis is not required to confirm a diagnosis of androgen deficiency. At the same time, blood count (androgen-deficiency anemia? polyglobulism?) and prostate-specific androgen (PSA) are determined. This minimum diagnosis can be extended to include further parameters depending on the range of symptoms and the patient's interest (☞ Figure 5.6). During therapy, changes can be expected in lipid metabolism parameters (total cholesterol, HDL and LDL cholesterol), the adipocyte hormone leptin and bone metabolism parameters (pyridinoline, desoxypyridinoline). However, these parameters as well as other hormone analyses (e.g. estradiol, DHT, DHEA sulfate) are not required to reach a decision on therapeutic indication.

	Symptoms	none (1)	slight (2)	moderate (3)	severe (4)	very severe (5)
(1)	Deterioration in general health (State of health, subjective well-being)	❍	❍	❍	❍	❍
(2)	Joint and muscle symptoms (Lumbago, arthralgia, neuralgia, back pain)	❍	❍	❍	❍	❍
(3)	Severe sweating (Unexpected/sudden outbreaks of sweating, hot flushes not induced by exercise)	❍	❍	❍	❍	❍
(4)	Sleep disorders (Problems going to sleep, staying asleep, waking too early or waking up tired, sleeping badly, sleeplessness)	❍	❍	❍	❍	❍
(5)	Increased need to sleep, frequently tired	❍	❍	❍	❍	❍
(6)	Irritability (Aggression, lose temper easily over trivial things, out of sorts)	❍	❍	❍	❍	❍
(7)	Nervousness (Inwardly tense, agitated, unable to keep still)	❍	❍	❍	❍	❍
(8)	Anxiety (Panic)	❍	❍	❍	❍	❍
(9)	Physical exhaustion/decreased strength (General decrease in vitality, decrease in activity, lack of enthusiasm, feeling of being able to do less, achieving less, having to drive oneself to do things)	❍	❍	❍	❍	❍
(10)	Decrease in muscle power (Feeling weak)	❍	❍	❍	❍	❍
(11)	Depression (Despondent, sad, cry easily, no drive, mood swings, feeling of pointlessness)	❍	❍	❍	❍	❍
(12)	Feeling best part of life is over	❍	❍	❍	❍	❍
(13)	Feeling discouraged	❍	❍	❍	❍	❍
(14)	Reduced growth of facial hair	❍	❍	❍	❍	❍
(15)	Decreased potency	❍	❍	❍	❍	❍
(16)	Decrease in number of morning erections	❍	❍	❍	❍	❍
(17)	Decreased libido (Sex less fun, hardly want intercourse)	❍	❍	❍	❍	❍

Table 5.5: The Aging Males' Symptoms (AMS) scale: a questionnaire about subjective estimation of symptoms and feelings of ill-health. Correlation with biochemical or objective criteria was not investigated. However, the severity of symptoms correlates with subjective estimate of androgen deficiency by an experienced investigator[149].

Optional diagnostic program

Case history
- Sexual history
- Backache, fractures
- Decrease in height
- Neurasthenic fatigue
- Tiredness

Technical investigations
- Bone density
- Bioimpedance (body composition)
- Dynamometer (measurement of strength)
- Rectal ultrasound

Examination
- Hunch-back
- Hip-waist ratio
- Muscle atrophy, muscle strength

Laboratory parameters
- Testosterone, LH, FSH, SHBG, free testosterone, prolactin, estradiol
- Blood count
- Liver values
- Bone metabolism parameters e.g. deoxypyridinoline
- Electrolytes
- Lipid metabolism (cholesterol, HDL, LDL)

Figure 5.6: Extended diagnostic program to supplement minimal program. Further investigations may be necessary depending on the clinical picture and interest of the patient.

If there is bone pain or pathological fracture, bone density should be measured (osteodensitometry) to estimate mineralization and diagnose osteopenia or osteoporosis. If dual photon absorptiometry (DEXA) is used, body composition can be calculated at the same time. If necessary, muscle strength can be quantitatively determined using a handheld dynamometer. Transrectal sonography of the prostate is useful for measuring volume of the prostate and excluding any focal changes, which have to be eliminated before testosterone therapy is started. The national and international guidelines require that digital rectal examination of the prostate and PSA determination be carried out before therapy[265,386].

5.4. Contraindications and monitoring of androgen substitution in older men

Carcinoma of the prostate is an absolute contraindication to testosterone substitution in men, and a digital rectal examination of the prostate must be carried out before treatment starts, in addition to PSA determination. A suspect finding must be clarified, possibly by taking a punch biopsy. If the findings are unclear or fail to exclude a malignancy, as a precaution testosterone substitution should not be given.

Caution should be exercised when administering testosterone to men with benign hyperplasia of the prostate (BPH) or significant symptoms (e.g. according to the International Prostate Symptom Score (IPSS)). Studies carried out to date (see below) have found no evidence of increasing prostate enlargement under testosterone substitution; however, patients with a high IPSS were not entered in the studies.

Polyglobulism (increased hematocrit) is another contraindication. Testosterone stimulates erythropoiesis and increases hematocrit still further. This impairs rheology, promotes thromboses and emboli and increases cardiovascular risk. Men with polyglobulism should be given a general examination and be excluded from therapy. Mammary carcinoma in men is another contraindication, as growth can be hormone-dependent. Severe heart failure is a relative contraindication. Testosterone substitution might increase the preload on the heart owing to slight water retention, and thus promote decompensation.

In rare cases, testosterone substitution may induce sleep apnea syndrome in predisposed men (e.g. those with pronounced obesity and chronic obstructive pulmonary disease). Patients affected with this condition often have an excessive increase in hematocrit (over 50 %) and polyglobulism. When testosterone therapy is discontinued, the sleep apnea syndrome subsides[249,328]. Nevertheless, caution is advised in patients with predisposing factors. They should be questioned about their sleep patterns during androgen substitution. Increasing tiredness during the day may be an indication of sleep apnea syndrome.

Available data show that it is safe to use testosterone in older men provided that the contraindications are taken into account (☞ Figure 5.7). Other possible side-effects include a feeling of tension in the breasts. However, the effects of long-term substitution on the development of prostate cancer have not yet been fully studied.

The dose is adjusted 6-12 weeks after the start of therapy. In transdermal substitution with Andro-/Testogel®, total testosterone should remain within the normal range, and, in testosterone enanthate administration ('3-weekly syringe'), total testosterone should be in the middle-normal range 7-10 days after injection and decrease below the lower

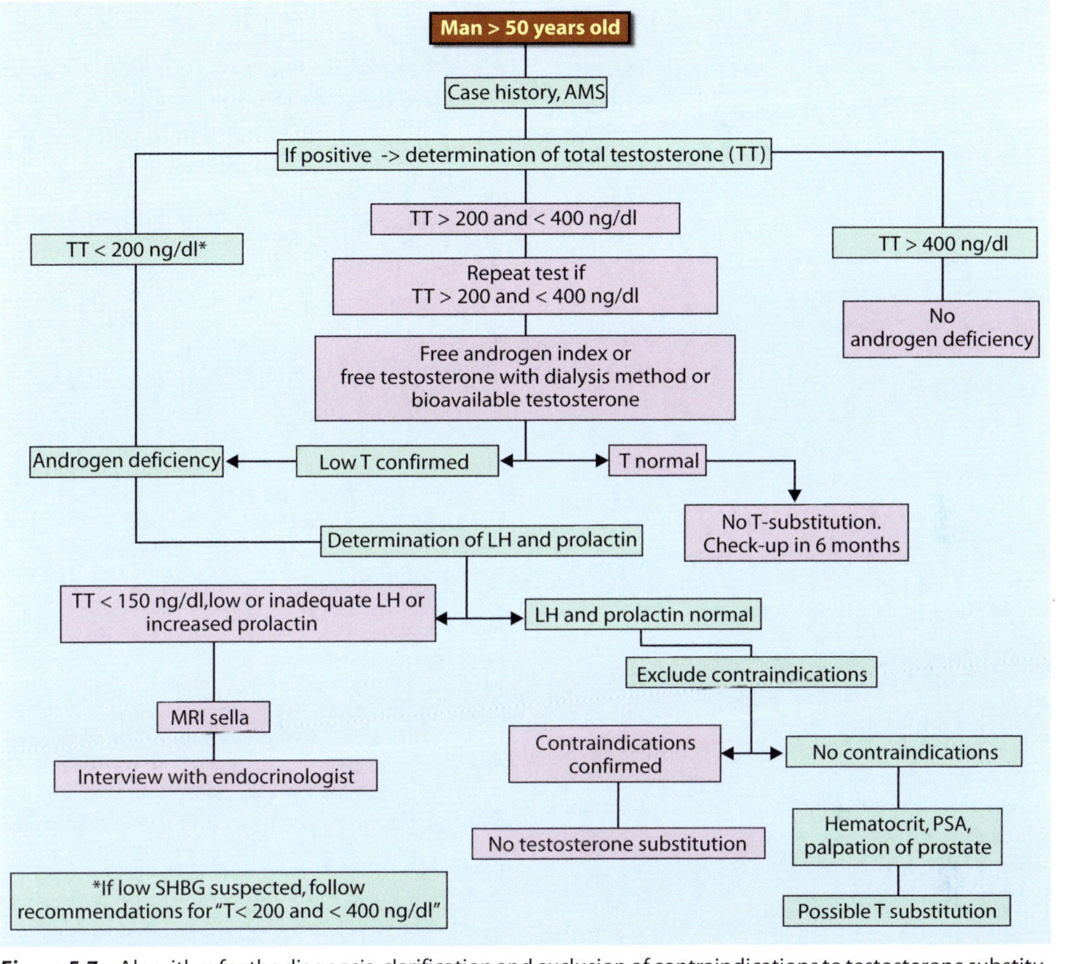

Figure 5.7: Algorithm for the diagnosis, clarification and exclusion of contraindications to testosterone substitution in late-onset hypogonadism according to the 2001 Consensus Conference of the American Endocrine Society[386]. AMS, Aging Males' Symptoms Score (☞ Table 5.5).

normal limit just before the next injection. Higher values signal an accumulation. With testosterone patches, the aim is to achieve a total testosterone level within the normal range 3-10 h after administration.

Therapy is initially monitored every 3 months by determining prostate-specific antigen (PSA) and palpating the prostate or performing transrectal sonography[171,266]. Blood count is also taken and lipid metabolism measured (including HDL). If PSA increases by more than 1.5 ng/ml within one year or there is an annual increase of more than 0.75 ng/ml in the consecutive 2 years or the PSA increases above the normal limit, a urological investigation should always be carried out. Liver toxicity rarely develops when using the recommended preparations. Osteodensitometry is performed at intervals of several years to monitor for signs of osteopenia. Because they are included in a structured monitoring program, men being treated with testosterone are probably better off in terms of the risk of developing prostate carcinoma than men not receiving treatment, who do not tend to undergo prophylactic screening (☞ Figure 5.8).

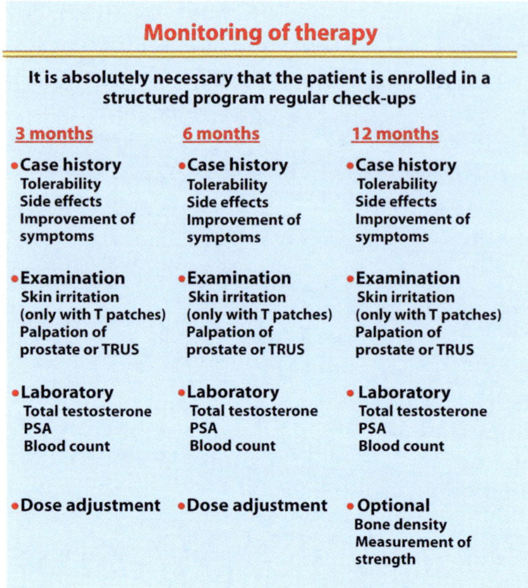

Figure 5.8: Monitoring success of testosterone substitution in late-onset hypogonadism[386].

5.5. Effects of testosterone substitution in older men

Testosterone substitution is expected to have a beneficial effect on bone metabolism, musculature, erythropoiesis, libido, sexual satisfaction and general mood in men with late-onset hypogonadism (☞ Figure 5.9). However, there is limited experience with testosterone substitution in older men, based mainly on studies with treatment times of less than 36 months. Therefore, testosterone substitution is not considered to be a standard treatment for late-onset hypogonadism.

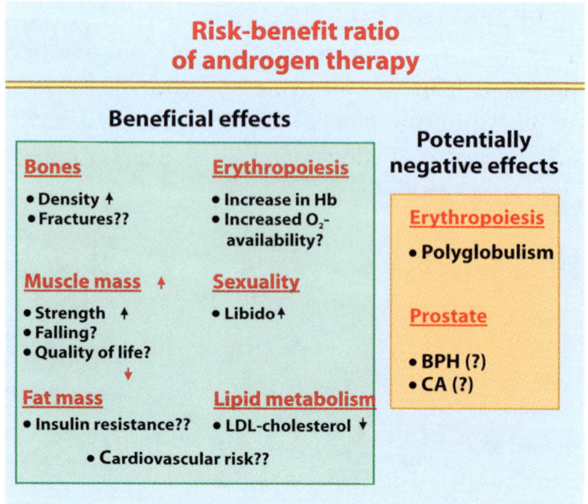

Figure 5.9: Overview of risk–benefit analysis. Although beneficial effects on bone metabolism have been verified it is not clear whether fractures are prevented. Muscle mass and strength increase although it is not known whether this leads to a functional benefit with improved quality of life and reduced incidence of falling. Adipose tissue is reduced. Influence on insulin resistance is unclear, as is overall effect on the cardiovascular system. Any positive benefits from increased ability to transport oxygen are unknown. Potential risks include polyglobulism and disorders of the prostate.

5.5.1. Influence on bone metabolism

Androgens have an osteoanabolic effect. However, 17β-estradiol is more important to bone metabolism in men than testosterone. Because the enzyme aromatase is present in osteoblasts, these are capable of converting androgens to estrogens[382].

Studies of the effect of testosterone substitution on bone metabolism in older men show a reduction in bone loss and reduced excretion of bone degradation parameters (hydroxyproline, pyridinoline)[168,207,313,385]. Four studies conducted over periods of between 12 and 36 months evaluated bone density and revealed an increase in spinal bone density[207,313,368]. In the largest randomized, placebo-controlled study, a bone density increase of around 4.2 ± 0.8 % was measured over a period of 36 months in the group treated with scrotal testosterone patches. After transdermal application of testosterone gel, a significant increase in bone density was observed after only 6 months of treatment[421]. Men with a significantly low initial serum testosterone level (< 200 ng/dl) benefited significantly more (by almost 6 %) than men with an only slightly reduced serum testosterone level[368]. Even men with corticosteroid-induced osteoporosis benefited from testosterone substitution and an increase in bone density[313], especially important since testosterone deficiency is particularly prevalent in this group of patients.

Even though there is still no evidence of testosterone substitution reducing fracture rate, the available data indicate a favorable effect on bone metabolism, and therefore, in addition to the usual basic osteoporosis therapy of vitamin D (cholecalciferol) and calcium, testosterone substitution can be given to older men with osteopenia/osteoporosis and androgen deficiency. There are no studies of combined treatment with bisphosphonates.

5.5.2. Influence on musculature and body composition

Numerous studies in healthy volunteers[41,438], body-builders, men with primary or secondary hypogonadism[42] and men with late-onset hypogonadism show androgens to increase lean body mass[39,385]. A considerable proportion of increased lean body mass is in muscle, and therefore muscle strength increases [41,42,60,134,301,420]. Even short-term testosterone therapy in older men causes a slight increase in muscle mass of around 3-8 %, resulting in a functional increase in muscle strength (☞ Table 5.6)[271,385,399]. However, not all studies were able to show these effects, and the critical endpoints of quality of life, morbidity and mortality were not investigated, even in the longer-term studies. Nevertheless, it is obviously still possible to reverse the process of catabolism even in older men. Since factors other than androgen deficiency contribute to muscle atrophy in old age, e.g. deficiency of growth hormone (GH) and insulin-like growth factor I (IGF-1), the proven interaction of testosterone with GH and IGF-1 has a particularly beneficial effect[200,399]. There is usually a 6-17 % reduction in fat mass, mirroring the increase in lean body mass. Testosterone substitution in older hypogonadal men maintains body composition at the level of younger eugonadal men[320]. However, the clinical significance of leptin suppression by androgens described above is still not clear in terms of its effects on body composition[179]. The alteration in fat mass is approximately equivalent to results from testosterone substitution in young men with hypogonadism. However, the increase in muscle mass is not as significant in older men as in younger men.

5.5.3. Influence on lipid metabolism

Whilst testosterone appears to have an adverse effect on lipid metabolism in young men with classical forms of hypogonadism as well as sportsmen who abuse androgens[180], testosterone seems to exert a beneficial effect on the lipid profile of older men[39]. Numerous smaller studies unanimously show a slight fall in total cholesterol and LDL cholesterol without any change in HDL cholesterol[143,243,271,385,441]. In contrast, a larger randomized study in 108 men over 65 years of age treated with testosterone patches or placebo was unable to show any significant change in LDL, HDL or total cholesterol over a 36-month period of treatment[368]. It is not known whether a reduction in LDL cholesterol results in a reduced incidence of cardiovascular disease in the long term. No end point studies have been carried out of the cardiovascular risk associated with testosterone substitution.

n	Age (years)	Therapy	Duration (months)	Fat mass	Lean mass	Strength	Source
12	68 ± 2	TE 100 mg/week	3	unchanged	+3.2 %	unchanged	385
10	57 ± 2	125 mg T-gel/day	9	-6.4 %	+2.4 %	-	243
8	78 ± 2	TE 200 mg/2 weeks	3	-4 %	-	+13 %	271
6	67 ± 2	TE 100 mg/week	1	-	-	+5–20 %	399
15	68 ± 6	TC 200 mg/2 weeks	12	-7.9 %	-	+10 %	355
54	73 ± 6	1 scrotal patch/day	36	-3 kg	+1.9 kg	unchanged	369

Table 5.6: Influence of testosterone substitution on fat mass, lean mass and muscle strength of older men. n = number of participants; TE = testosterone enanthate; T = testosterone; TC = testosterone cypionate.

On the other hand, a randomized controlled study in 227 hypogonadal men receiving testosterone gel therapy showed a significant reduction in total cholesterol and a clear reduction in LDL cholesterol[420].

The treatment of obese men with androgens reduces accumulation of triglycerides in the subcutaneous adipose tissue of the abdomen where it reduces activity of the lipoprotein lipase, although it does not have the same effect in the subcutis of the extremities[245,246]. Selective reduction in abdominal adipose tissue exerts a beneficial effect on carbohydrate metabolism and blood pressure. The clinical significance of suppression of leptin by androgens with regard to body composition and lipid metabolism is still unclear[179], but it may be involved in the reduction of fat mass as a whole. Therefore, carbohydrate metabolism benefits from testosterone substitution. Placebo-controlled studies in obese men over 45 years of age unanimously showed an improvement in insulin sensitivity, fasting blood sugar and diastolic blood pressure during testosterone therapy[243,244]. In complete accordance with these findings is a recent pilot trial with 48 men with diabetes mellitus type 2[55]. The 24 men receiving oral testosterone undecanoate achieved a significant reduction in body weight, waist-hip ratio, and body fat (-5.65%) with a simultaneous improvement of metabolic control. The mean glycosylated hemoglobin (HbAlc) decreased from 10.4 to 8.6%. In contrast, the control group did not show any significant changes[55]. However, these studies are still awaiting confirmation. On the other hand, low concentrations of testosterone are associated with an increased risk of developing type II diabetes mellitus, and diabetics show reduced levels of testosterone[142,372]. In addition, the Rotterdam study showed a link between low levels of testosterone in older men and an increased risk of developing atherosclerosis[144].

5.5.4. Influence on vascular reactivity

New experimental studies show testosterone to have an acute vasodilatory effect. Intracoronary infusion of testosterone during coronary angiography to men (61 ± 11 years old) with coronary heart disease (CHD) had an acute vasodilatory effect. It also increased blood flow[425]. Moreover, intravenous administration of testosterone to men with CHD immediately prior to exercise ergometry reduces myocardial ischemia during exercise. Electrocardiogram (ECG) changes take longer to develop and overall exercise time is extended[321,424]. Some time ago, a randomized double-blind study in 50 men (aged 35-71) with CHD showed that 8 weeks of treatment with 200 mg intramuscular testosterone cypionate per week significantly reduced ST segment depression in the ergometric test by 51 %, whereas placebo showed no effect[172]. In contrast, one study in 32 men showed that intravenous administration of testosterone had no effect on myocardial perfusion[389]. Some cardiovascular risk factors appear to benefit from testosterone substitution, however, studies which evaluate objective end-points such as cardiovascular events or mortality would be more persuasive.

5.5.5. Influence on erythropoiesis

Testosterone stimulates erythropoiesis directly by its effect on stem cell proliferation, and indirectly by increasing renal production of the hematopoietic growth factor erythropoietin. Testosterone substitution in young men with hypogonadism usually causes an increase in red cell count, hemoglobin concentration and hematocrit[190,420]. Erythropoiesis is also stimulated by testosterone substitution in older men: an increase in hematocrit of around 3-7 % has been observed[271,355,385]. The increased ability to transport oxygen may be associated with an improvement in physical capacity and individual organ function, although this has yet to be proven[190].

5.5.6. Influence on mental well-being and cognitive abilities

There are few reliable data on the effect of androgen therapy on mood and mental well-being in men with late-onset hypogonadism. Studies in young hypogonadal men showed an improvement in mood and emotional well-being during testosterone substitution[158,420]. This is confirmed by clinical experience in older men[243,385].

A recent clinical trial in men with refractory depression and low or borderline serum testosterone levels confirmed the clinical experience[308]. In a randomized, placebo-controlled study, subjects receiving testosterone gel for 8 weeks had significantly greater improvement in scores on the Hamilton depression rating scale (HAM) than subjects receiving placebo. Changes were noted on both the vegetative and affective subscales of the HAM. Furthermore, a significant improvement was also seen on the Clinical global impression severity scale, but not on the Beck depression inventory[308]. A significant improvement over placebo was also shown in another trial, where men with late-life major depression received testosterone[299]. Interestingly, men with late-onset depression responded considerably better to testosterone substitution than men with early onset depression. Thus, testosterone seems able to particularly improve depression in the elderly male.

A 3-month testosterone substitution treatment in older men caused an improvement in spatial thinking, but not verbal or visual memory, motor speed or cognitive flexibility[175]. Another study was completely unable to detect any cognitive effects[355], although testosterone did improve spatial and verbal memory in a third study[78]. Numerous studies indicate that sex steroids have a differentiated effect on organocerebral performance, and, among other things, effect spatial powers of imagination, linguistic ability and memory[9,40,79]. Nevertheless, the current instruments for recording these effects appear to be unsuitable.

Testosterone substitution is not expected to correct erectile dysfunction in men with late-onset hypogonadism. On the other hand, testosterone substitution increases sexual satisfaction in older men, and therefore therapy may be attempted if sexual appetite has reduced[143].

5.5.7. Influence on the prostate

There is a high incidence of benign hyperplasia of the prostate (BPH) and carcinoma of the prostate among older men. Androgens appear to play a significant role in the clinical expression of both diseases, and therefore the potentially adverse effect of androgen therapy in older men is causing great concern in specialist circles. Studies supporting this are not yet finished.

The frequency of benign hyperplasia of the prostate (BPH) rises from 30 % in 70-year-olds to almost 100 % in centenarians[92]. Although testosterone production reduces in the development stage of BPH, paradoxically, androgens, in particular the 5α-reduced testosterone metabolite dihydrotestosterone (DHT), play an important role in the development of BPH. BPH practically never develops in castrated or hypogonadal men, and reducing systemic concentrations of androgen or discontinuing therapy causes BPH to recede[119,310]. Nevertheless, there are no differences in serum androgen concentrations between men with and without BPH or carcinoma of the prostate, and therefore presumably the metabolism of androgens to DHT and estradiol is crucially important from a cellular viewpoint[221,233,256,310]. The most extensive analysis to date of all available epidemiological data found no evidence of a connection between sex steroids and carcinoma of the prostate[52].

Nevertheless, it seems unlikely that the administration of testosterone to older men with testosterone deficiency promotes the *de novo* development of malignant cells[162,354], but there is a risk that it might promote the growth of existing malignant cells[361]. Approximately 50 % of all men over 70 display preclinical microscopic foci of malignant cells in the prostate, although there are great ethnic and geographical differences in progression to clinically manifest carcinoma, and the majority of patients do not develop clinically apparent carcinoma of the prostate. Androgens appear to play a permissive role in the development of malignant cells; however, they do not induce malignancy[92,163,338,361]. It is not known whether androgens promote the progression of preclinical foci to clinically manifest prostate cancer. However, the fact that established prostate carcinoma is dependent on testosterone for growth has been documented, and is also exploited therapeutically.

Only a few studies of the effect of testosterone therapy on the prostate have been performed. Substitution of testosterone in men with hypogonadism stimulates prostate growth and normalizes serum PSA concentration and prostate size[32,271]. In a placebo-controlled study, the administration of 160 mg testosterone undecanoate per day to 11 eugonadal men aged between 40 and 65 caused a 12 % increase in prostate size and no change in PSA after 8 months[156]. In three out of six studies of the effect of testosterone therapy on the prostate of older men, there was a slight but significant increase in prostate-specific antigen (PSA) (☞ Table 5.7); however, these increases were not classified as clinically significant. In particular, the two studies lasting 2 and 3 years, respectively, did not reveal a higher incidence of carcinoma of the prostate,

n	Age in years	Medication	Duration (months)	PSA before therapy	PSA after therapy	Source
13	68 ± 2	TE 100 mg/week	3	2.1 ± 0.4	2.7 ± 0.5[a]	385
8	77 ± 2	TE 200 mg/2 weeks	3	1.7 ± 0.6	2.5 ± 0.8	271
6	67 ± 2	TE 100 mg/week	1	1.7 ± 0.6	2.1 ± 0.7[b]	399
15	68 ± 6	TC 200 mg/week	12	1.0 ± 0.2	1.9 ± 0.3	355
45	72 ± 2	TE 200 mg/2 weeks	24	1.3 ± 0.2	1.8 ± 0.2	143
54	73 ± 6	1 patch/day	36	1.9 ± 1.4	2.2 ± 1.8[c]	368

Table 5.7: Serum PSA concentration (in ng/dl) before and after therapy in older men on testosterone treatment. Doses given in mg; n = Number of patients; TE = testosterone enanthate; TC = testosterone cypionate (behaviour identical to TE); the Testoderm scrotal plaster was used. [a] $p < 0.01$; [b] $p < 0.05$; [c] $p < 0.005$.

BPH or other conditions of the prostate, compared with untreated men[143,368]. In a non-randomized, controlled 3-year observation of men with andropenia receiving testosterone therapy, only 7 % of the treated men (compared with 22 % of the untreated men) developed BPH or carcinoma of the prostate[143]. Surprisingly, 1 trial showed even a reduction in prostate volume, PSA and lower urinary tract symptoms (International prostate symptom score, IPSS) when substituting men with low or borderline serum testosterone levels with oral testosterone undecanoate for 3 months[296]. Although this study awaits confirmation, it adds further evidence to the studies, that prostatic complications are not necessarily encountered, when substituting elderly men with testosterone. Therefore, on the basis of data available to date, testosterone may be administered to older men provided that PSA is checked regularly and ultrasound images are taken of the prostate[265,386]. In the future, patients may be stratified on the basis of their PSA, to identify those at increased risk of developing tumors of the prostate and exclude them from testosterone substitution[434].

5.6. Differential therapy

Testosterone is indicated in andropenia if serum testosterone concentrations are repeatedly low, there are clinical signs of androgen deficiency and contraindications have been excluded. If serum testosterone is reduced but there are no clinical signs of androgen deficiency, then there is usually no reason for therapy as the patient will not derive any benefit from it. The same is true of the reverse situation, where clinical signs of androgen deficiency are present but testosterone levels are normal: there is no evidence that androgen administration will be of any benefit if testosterone levels are normal.

After making a detailed evaluation and weighing up the risks and benefits, testosterone can be assumed to have a beneficial effect on bone metabolism, muscle mass, muscle strength, sex-life and androgen-deficiency anemia[39]. Testosterone also improves functional capacity, increases bone mass, has a favorable effect on lipid metabolism, increases sexual satisfaction and improves quality of life. It is not known whether there is a beneficial effect on morbidity and mortality in the long term. On the other hand, testosterone has a potentially adverse effect on the prostate, although this is less significant than previously assumed according to the results of recent studies, and can be contained by asking the patient to undergo regular check-ups.

Accordingly, the pros and cons should be discussed with the patient on the basis of recorded findings and hormone status. If there are no contraindications to testosterone therapy and there are several disorders present which could be improved by testosterone, the decision to attempt therapy is an easy one, even if serum testosterone levels are only slightly reduced. However, if there are no clinical signs of androgen deficiency and no improvement in quality of life or symptoms can be expected, even if serum testosterone levels are low, restraint should be exercised. A 3-6-month trial with testosterone might help to reach a decision about long-term treatment. Symptoms could then be re-evaluated after this time. If symptoms improve during the trial period, therapy should be continued.

It is important that the patient returns for regular check-ups. These should initially take place every 3 months.

Initially, PSA and blood count are the most important parameters for monitoring potential side-effects. Serum testosterone level should also be checked, and the testosterone dose or injection interval adjusted if necessary. Later on, check-ups are spaced out and other parameters added. It is particularly important that all tests which led to the initial pathological findings are repeated. However, there is no sense in repeating osteodensitometry within a year because bone metabolism is very slow.

Almost all studies investigating testosterone therapy in andropenia used testosterone enanthate at the usual dosage of 200-250 mg intramuscularly every 2-3 weeks (☞ Table 4.4). Testosterone gel is particularly easy to administer. It is applied once a day, and the patient is not dependent on intramuscular injections. Skin tolerability is very good and serum testosterone levels steady, unlike with some injectable testosterone esters. When used in classical forms of hypogonadism, testosterone gel has proved to be as effective as the standard preparation testosterone enanthate[380,420]. An added advantage of testosterone gel is that therapy can be discontinued immediately if there are side-effects or prostate findings are unclear, whereas intramuscular injections stay in the system longer. For the same reason, testosterone pellets and injections which have a very long duration of effect (e.g. intramuscular testosterone undecanoate) are not as suitable for initiation of testosterone treatment in late-onset hypogonadism. Testosterone patches can be used as an alternative to the gel. However, these cause a high rate of skin irritations and are always visible, whereas the gel is absorbed by the skin and is therefore inconspicuous. The scrotal testosterone patch increases testosterone levels but also causes supraphysiological serum concentrations of dihydrotestosterone. It is not clear whether this has an adverse effect on the prostate in older men. Bioavailability of oral testosterone undecanoate is poor, and therefore its efficacy is very variable. Patients with late-onset hypogonadism are particularly aware of this, which has an adverse effect on compliance.

Estrogen substitution in men

6. Estrogen substitution in men

6.1. Estrogen production in men

Estrogens also have an important function in men. Estradiol is the most important biologically active estrogen; approximately 20 % of estradiol originates in the Leydig cells. Most circulating estradiol is formed in the peripheral tissues from aromatization of androgens, mainly testosterone, but some from adrenal androstenedione. Therefore, testosterone is a prohormone of estradiol. Activity of aromatase and hence conversion of androgens to estrogens increases with age and increase in body fat[408]. The amount of subcutaneous abdominal adipose tissue in particular, and to a lesser extent visceral adipose tissue, is an important predictor of estrogen levels. Serum levels of total estradiol are normally 2-3 ng/dl in men. Just like testosterone, estradiol is bound to SHBG, but its affinity is less than that of testosterone. With estradiol a distinction is also made between total estradiol and free, unbound, biologically active estradiol.

6.2. Age-related changes in estrogen levels

In young men there is a very good correlation between testosterone and estradiol and free testosterone and free estradiol. Therefore, as testosterone production decreases with age, one would expect serum estradiol levels also to decrease with age. This was not observed in the majority of studies set up to investigate this premise[131,408]. Whereas the largest study of this kind, the Massachusetts Male Aging Study, documented no age-associated change in estradiol levels in a cross-sectional study of 1700 men (although it recorded a decrease in testosterone and a reduction in estrone)[109,131], a small German study in 36 elderly men (67-93 years of age) actually detected higher estrogen levels than in younger men[306]. In contrast, a slight reduction in serum estradiol with increasing age was recently described. Men above the age of 60 had 30 % lower estradiol levels than men under 30[232]. This was also verified by a Dutch study[401]. Therefore, it is still not proven that men suffer a fall in estrogen levels with increasing age. Presumably, as the activity of aromatase increases with age, this compensates for the substrate deficiency due to falling levels of testosterone, and therefore estrogen production is maintained. Nevertheless, serum levels of free estradiol decrease with age, because SHBG concentrations increase with age[232,401]. It is not clear whether slim men with late-onset hypogonadism and correspondingly low levels of aromatase are more likely to suffer a deficiency of free estradiol.

Estrogen deficiency occurs only without simultaneous testosterone deficiency, in congenital aromatase deficiency. To date, this condition has been described in only four men worldwide. It is caused by a mutation in the aromatase gene. The clinical picture of these patients is characterized by severe osteoporosis and fractures, below the age of 30, and open epiphyses with growth in height continuing into adulthood[71,97,152,268].

6.3. Physiological significance of estrogens in men

Estradiol levels in young men are comparable with levels in young women in the early follicular phase, and significantly greater than those of postmenopausal women. Both types of estrogen receptor are expressed in many tissues in men. Hence, estrogen receptor α is found in the anterior lobe of the pituitary, the testes, liver, kidneys, bones and brain. Estrogen receptor β is found in bones, cartilage, gastrointestinal tract, thyroid, prostate, skin and urinary bladder. Although much is known about estradiol production and target tissues, not enough is known about the significance of estradiol in man. Nevertheless, our knowledge of estrogens and their mechanisms of action has grown enormously in recent years (☞ Chapter 2.2.4.4.). The following have contributed to this: discovery of the two estrogen receptors α and β (ERα, ERβ), discovery of men with inactivating mutations of the ERα or aromatase enzyme gene and the development of mouse models with inactivating estrogen receptors α or β (estrogen receptor knock-out, ERKO).

It has been known for some time that estrogens are responsible for negative feedback to the gonadotropins FSH and LH[113,114]. Estradiol reduces pulse height as well as pulse frequency of LH, and thus exerts a regulatory function for androgen biosynthesis in the testes via negative feedback to the hypophysis (☞ Figure 1.2). Paradoxically, estrogens play a critical role in masculinization of the CNS during the prenatal development stage[128,129,229]. In adult men too, estrogens appear to stimulate libido and sexual activity. Treating an aromatase-deficient man with estrogens will increase libido, frequency of sexual fantasies, masturbation and sexual intercourse[72]. ERKOαβ mice are completely disinterested in sex[288], and therefore estrogens are obviously an important stimulus of male sexuality.

In addition, estrogens trigger growth in puberty by increasing the rate of secretion of growth hormone by increasing pulse amplitude[250,397].

Estrogens are also important for male fertility. The Sertoli cells and Leydig cells, epididymis and testicular efferent ducts express estrogen receptors, as do the gametes themselves. ERKOα mice are infertile owing to a disturbance in the functioning of the epididymis and the rete testis. However, ERKOβ mice are not[288]. Some of the very few men with inactivating estrogen receptor α or aromatase deficiency also have disorders of testicular function[71,152,268,365]. However, the most striking characteristic of these patients is the disorder of bone metabolism. The lack of estradiol effect causes severe osteoporosis and excessive growth due to open epiphyses. Growth continues even into adulthood. In three patients with aromatase deficiency, the administration of estradiol led to closure of the epiphyses and a sharp increase in bone mass[71,152,268]. Therefore, estradiol, and not testosterone as previously thought, is the critical sex steroid for bone metabolism. Aromatase is formed in osteoblasts, and osteoclasts so the requirement for estrogens can be met by local production from testosterone[64,174,345]. A link was recently found between serum estradiol levels and bone density in older men[401]. Therefore, increasing the level of estrogen (e.g. by testosterone substitution) may have a beneficial effect on bone mass in older men with osteoporosis.

There are a number of studies of the effect of estrogens on lipid metabolism, endothelial function, vascular musculature and isolated aspects of cardiac function and cardiac output. The majority of these studies were carried out in women, and mostly show beneficial effects (of estrogen) on the surrogate parameters morbidity and mortality. It was recently shown in healthy older men that short-term oral administration (9 weeks) of up to 2 mg estradiol reduces levels of homocysteine, fibrinogen and plasminogen activator inhibitor-1 (PAI-1), and has a beneficial effect on lipid profile without increasing markers of thrombosis risk[124]. In contrast, a 4-week treatment in men with hypogonadism using lower doses of conjugated estrogens (0.3 mg per day) had no effect on lipid metabolism[292]. Lipoprotein(a) also decreases during estrogen therapy[154]. Information about the effect of estrogen on vascular function is contradictory. Whereas one study showed no reduction in acetylcholine-induced coronary vasoconstriction[81], other studies clearly demonstrated the vasodilatory effects of acutely administered intravenous conjugated estrogens after acetylcholine or cold stress in men with coronary heart disease or following a heart transplant[130,314,315]. These effects appear to be triggered by non-genomic mechanisms which do not appear to require ERα, since sublingual administration of estradiol to one patient with an ERα mutation produced rapid vasodilatation[374]. However, as in studies of the ERKO mouse, there was no blood flow-induced vasodilatation, and therefore presumably this effect is mediated by ERα[325]. Moreover, there appears to be a gender-specific distribution of ER in the smooth vessel musculature. Whereas more ERα than ERβ is found in women, equal levels of both are found in men[155]. It is not clear whether these gender differences are relevant or what the significance of ER on cardiomyocytes might be[138]. Nevertheless, more ERβ is expressed in vascular lesions[236].

Despite the fact that estrogens have a beneficial effect on vasodilatation and lipid parameters, estrogens can increase the risk of thromboembolism in women, as shown in the Heart and Estrogen/progestin Replacement Study (HERS)[159] and some studies of the contraceptive pill, and in men, as shown by the few clinical studies of estrogens in men[83,85,290,335]. Venous thromboemboli are an important complication of high-dose hormone therapy in male-female transsexuals[403]. Moreover, one case-control study recorded higher estradiol levels in men with myocardial infarction than in men with comparable cardiovascular disease but no infarction[303-305]. Estradiol might possibly promote coronary occlusion.

To date, the effect of estrogens on the prostate is not well understood. Both ERs are expressed in the prostate. ERα has been identified only in stroma cells, ERβ has been identified in stroma cells and epithelium[162,354]. The function of estrogen on the prostate is unclear. There were no conspicuous pathological findings in the prostate of ERKOαβ mice, although ERKOβ mice suffered from hyperplasia of the prostate, indicating an antiproliferative or proapoptotic effect of estrogens[162]. Moreover, the ERs are regulated differently. Therefore, normal basal cells reveal ERβ; however, severely dysplastic cells do not, which means the majority of moderately differentiated invasive prostate carcinomas[354]. The numbers of ERβ decrease as the Gleason score increases, but ERβ is present in metastasizing tumors which are not dependent on androgens[354]. This might mean that estrogens exert autonomous, androgen-independent effects on the proliferation behavior and differentiation of prostate cells. Finally, it is not clear whether estrogens inhibit or promote growth.

6.4. Clinical studies on estrogen therapy in men

Since 1966, against a background of increasing mortality from cardiovascular events, 8341 men with myocardial infarction have been treated with 2.5 mg conjugated estrogens ($n = 1101$), 5 mg conjugated estrogens ($n = 1119$) or placebo ($n = 2789$) to prevent a second heart attack and reduce morbidity rate to the significantly lower one of women. However, the 5 mg estrogen arm was discontinued as early as 1970, as the rate of non-fatal myocardial infarction was twice as high as in the placebo group. In addition, deep vein thrombosis was twice as high and pulmonary embolism four times as high in the 5 mg estrogen group than in the placebo group[83]. Three years later, the 2.5-mg estrogen arm was also stopped. The incidence of cardiovascular events and death had not decreased, compared with placebo, and in fact deep vein thrombosis and pulmonary embolism events were almost twice as high[85]. In retrospect, these disappointing results should have been foreseen. In 1969, one study carried out by the Veterans Administration in a total of 570 men with myocardial infarction was unable to show a beneficial effect from 1.25 mg conjugated estrogens. The mortality rate of the estrogen group was slightly increased, compared with the placebo group, although the increase was not significant[335]. Moreover, one British study using 200 µg ethinylestradiol even showed an increase in mortality, compared with placebo (13 out of 50 compared with 10 out of 50)[290]. These data contradicted those from a smaller study in which a reduction in mortality was observed with conjugated estrogens (1.25-2.5 mg)[248]. However, this study also included an unspecified number of men who had discontinued estrogen therapy after only 2.5 months but who continued to be recorded in the estrogen group for subsequent years. Hence, these data cannot be evaluated. No more large studies of the effect of estrogens on cardiovascular mortality have been carried out since this one; in fact, ethics committees would find it difficult to approve such studies.

Urological studies in which men with carcinoma of the prostate are treated with high-dose estrogens to suppress testosterone have provided new findings. These studies also show increased cardiovascular mortality in the estrogen groups compared with other treatment modalities[151,260].

As a whole, it can be concluded from the clinical studies that high doses of estrogens increase mortality rather than reduce it. It is not known whether lower doses have a different effect as this has not been investigated. It is not admissible to transfer findings from clinical studies in women to men, since, first, treatment with hormones from the opposite sex generally causes other effects, and, second, particularly in men, isolated administration of estrogens changes the equilibrium of androgens to estrogens and the effects of a relative reduction in androgens are unclear. Finally, these data are in accord with the results of prospective, randomized studies of hormone replacement therapy in postmenopausal women (Women's Health Initiative (WHI) -HERS), which also showed no reduction in cardiovascular mortality with estrogen substitution, despite the observation that most effects on surrogate parameters were beneficial[159]. Obviously, we do not yet understand how to use estrogens in such a way that the beneficial effects on surrogate parameters are also realized. It might be an issue of dosing or selecting patients according to risk factors. In any case, uncontrolled field tests on men outside a study context are totally unsuitable for resolving these issues.

6.5. Side-effects of estrogen therapy in men

In addition to the increased risk of cardiovascular mortality, men on estrogen therapy usually develop gynecomastia, which can be disturbing. This effect is dependent on dose; the incidence in larger clinical studies was 40-98 % (!)[83,85,290,335]. In addition, estrogen therapy can also be expected to reduce serum testosterone by suppressing gonadotropins. This can have an adverse effect on potency and fertility. It is conceivable that the lack of anabolic effect from testosterone encourages muscle atrophy and hence age-related sarcopenia, but this has not been investigated. Not least, it is not known what effect estrogens have on the development or progress of diseases of the prostate (☞ Chapter 6.3.).

6.6. Concluding remarks on estrogen therapy in men

At the present time, it is strongly adviseable not to treat men with estrogen. Clinical studies clearly show adverse effects on cardiovascular mortality, predominantly from thromboembolic events. This cannot be offset by more recent findings relating to beneficial effects on surrogate parameters (lipid metabolism, endothelial function, vascular reactivity). There is currently no evidence to show that estrogen therapy is any more beneficial than testosterone therapy in older men. The side-effects have not been fully characterized (e.g. gynecomastia, impotence, infertility, diseases of the prostate). Moreover, it is not clear whether older men actually suffer a clinically significant estradiol deficit. Any deficiency is more likely to be due to the decreasing production of testosterone. Hence, there is no rational basis for treating the symptoms of old age in men with estrogen. Numerous organs in men are dependent on a balanced ratio of androgens to estrogens for proper functioning. Therefore, any requirement for hormone therapy in the context of late-onset hypogonadism should be compensated with testosterone, which is metabolized to estrogens, increasing levels of 17β-estradiol in the serum. Data for evaluating the risk-benefit ratio are available in Chapter 5.[184,191,196].

Case reports

7. Case reports

The immediate treatment of a patient is always the primary concern of a doctor. Therefore, this chapter has been specially devoted to selected case histories from the out-patient endocrinology/andrology clinic. This will show how the diagnosis, treatment and long-term care of patients with androgen deficiency are managed in practice.

From experience the reader will gain more information, and learn more if he determines the diagnostic procedure, diagnosis and therapeutic consequences for himself. Therefore, diagnosis and suggestions for therapy should not be read in advance.

Case report 1

Case history

A 34-year-old man with bilateral orchiectomy (right side, 4 years ago and left side, 3 years ago) due to seminoma, recurrence-free since then

? What laboratory parameters would you measure?

Examination findings

184 cm, 79 kg, normal general medical examination

Biochemical findings

The following were normal or negative:
ESR, TU marker
The following were low or pathological:
hemoglobin 11.8 g/dl, hematocrit 37 %,
PSA not detected

? What hormone parameters would you measure?

Hormone analysis

		Normal range
Testosterone	1.7	12-30 nmol/l
LH	22.4	1-10 U/l
FSH	45.8	1-10 U/l

? What additional tests would you recommend?

Additional tests

Osteodensitometry: normal

? What treatment would you give?

Diagnosis/evaluation

Acquired anorchidism with primary hypogonadism

Therapy

Life-long testosterone substitution, e.g. with 250 mg testosterone enanthate i.m. every 3 weeks or 1000 mg testosterone undecanoate i.m. every 12 weeks

? Why is the patient anemic?

Simple case to start with. Patient with primary hypogonadism as a result of acquired anorchidism which requires life-long substitution.

✓ **Question 1:** Blood count due to androgen-deficiency anemia, PSA before testosterone substitution, possibly liver parameters (GOT, GPT, bilirubin).
✓ **Question 2:** Testosterone, LH und FSH are sufficient to verify hypogonadism before start of therapy. SHBG and estradiol provide no additional information.
✓ **Question 3:** Bone density due to osteoporosis risk caused by androgen deficiency.
✓ **Question 4:** Permanent substitution possibly with long-acting preparations. Testosterone gel or testosterone patches would also be a possibility but are exceedingly expensive for life-long treatment of a young patient and there is no proof that they are any better.
✓ **Question 5:** The anemia is due to androgen deficiency and is corrected by substitution of testosterone.

Case report 2

Case history

A 34-year-old man, referred as unable to have children, sexual potency apparently good

? What tests would you order?

Examination findings

182 cm, 86 kg, sparse secondary hair, testis volume 3 ml both sides, no gynecomastia

Sperm test

Azoospermia

Hormone analysis

? What is your suspected diagnosis and how would you prove this?

		Normal range
Testosterone	12.3	12-30 nmol/l
LH	17.1	1-10 U/l
FSH	24.3	1-10 U/l

Additional tests

Chromosome analysis: 47,XXY

Osteodensitometry: osteoporosis, bone density around 2.8 SD lower than normal values for age

? Additional tests?

Diagnosis/evaluation

Klinefelter's syndrome with secondary osteoporosis

Therapy

250 mg testosterone enanthate i.m. every 3 weeks or 1000 mg testosterone undecanoate i.m. every 12 weeks.

Recommend adoption, alternatively donor insemination

? What treatment would you give?

Patients with Klinefelter's syndrome typically consult a doctor with infertility or signs of androgen deficiency.

✓ **Question 1:** Semen analysis as patient consulting because unable to have children.

✓ **Question 2:** The combination of small testes, azoospermia and hypergonadotropic hypogonadism is typical of Klinefelter's syndrome and should always lead to chromosome analysis to confirm the diagnosis.

✓ **Question 3:** Bone density should be one of the additional tests. This should be obtained before therapy to obtain a starting value. The bone density test should be repeated every 1–2 years to ascertain whether osteoporosis is being prevented.

✓ **Question 4:** Life-long substitution with injectable testosterone esters. Testosterone gel or testosterone patches would also be a possibility but are exceedingly expensive for life-long treatment of a young patient and there is no proof that they are any better.

Case report 3

Case history

A 35-year-old man with Hodgkin's lymphoma, 7 years ago, for which given chemotherapy and radiotherapy (including pelvis and both sides of groin), recurrence-free since then.
Complains of lack of libido and impotence, backache, neurasthenic fatigue and tiredness

? What laboratory parameters would you measure?

Examination findings

184 cm, 79 kg, normal general medical examination, scar where lymphoma extirpated, left groin, testis volume right side 10 ml, left side 9 ml, slightly solid consistency, prostate normal

Biochemical findings

		Normal range
ESR	8	< 20
Hemoglobin	14.1	14-18 g/dl
Hematocrit	42	42-50 %
Leukocyte	4500	< 11000/µl
PSA	0.31	< 4 ng/dl

? What is causing the hypogonadism?

Hormone analysis

		Normal range
Testosterone	7.9	12-30 nmol/l
LH	12.1	1-10 U/l
FSH	19.8	1-10 U/l

? What additional tests would you order?

Diagnosis/evaluation

Primary hypogonadism due to radiotherapy and chemotherapy

? Would you give treatment?

Therapy

250 mg testosterone enanthate i.m. every 3 weeks or 1000 mg testosterone undecanoate i.m. every 12 weeks

Depending on type and dose, radiotherapy and chemotherapy can cause damage to the testicles. This patient obviously also has irreversible infertility that cannot be treated as the small testes and very high FSH indicate.

✓ **Question 1:** ESR (recurrence?), blood count (androgen-deficiency anemia? bone marrow damage?), PSA before possible testosterone therapy, possibly liver values.

✓ **Question 2:** Chemotherapy and radiotherapy.

✓ **Question 3:** Bone density, as high risk of osteoporosis (due to radiotherapy).

✓ **Question 4:** Therapy as for other patients with primary hypogonadism with injectable testosterone esters. Depending on results of osteodensitometry, additional treatment should be given for osteoporosis which is the cause of the backache.

Case report 4

Case history

A 55-year-old man, erectile dysfunction, loss of libido, tiredness, decreasing physical capacity, when questioned complains of increasing headaches, used to shave more frequently

? Which hormones would you analyze?

Examination findings

172 cm, 86 kg, sparse secondary hair, testis volume 12 ml left, 13 ml right, slight gynecomastia, no galactorrhea

Biochemical findings

		Normal range
Hemoglobin	12.4	14-18 g/dl
PSA	0.3	< 4 ng/dl

? What additional tests would you order?

Hormone analysis

		Normal range
Testosterone	4.9	12-30 nmol/l
LH	0.8	1-10 U/l
FSH	1.1	1-10 U/l
Prolactin	3750	< 20 ng/dl

? What treatment would you suggest?

Additional tests

LHRH test: insufficient increase in LH and FSH

MRI of hypophysis: 3.5 cm intra-, supra- and parasellar mass

Eye consultation: bitemporal hemianopia starting

Endocrinology: thyroid and adrenocorticotropic hypophyseal function absent

? What must you consider with hypophyseal tumors?

Diagnosis/evaluation

Macroprolactinoma with secondary hypogonadism, secondary hypothyroidism and secondary adrenal insufficiency

Therapy

250 mg testosterone enanthate every 3 weeks, dopamine agonist, e.g. bromocriptine, quinagolide, cabergoline, substitution with levothyroxine and cortisone

Patient with insufficiency of the anterior lobe of the hypophysis underlying the main typical symptom of loss of libido and potency. Approximately 80 % of all patients with hypophyseal adenoma suffer from increasing headaches. Then you have to consider determining prolactin, especially if there is gynecomastia. If the prolactin is very high this will determine the course of treatment. The urologist must then consider whether further tests are required, and if necessary consult an endocrinologist to co-ordinate subsequent diagnosis. With microadenoma (rare in men) one can wait to see whether dopamine agonist therapy is successful before administering testosterone. In these patients testosterone often normalizes after 1–2 months

✓ **Question 1:** Prolactin. The headaches and gynecomastia combined with the clinical evidence of androgen deficiency should lead one to consider hypophyseal adenoma.

✓ **Question 2:** MRI of sella, hypophyseal function tests, if necessary enlist help of endocrinologist.

✓ **Question 3:** Testosterone substitution; dopamine agonists are therapy of choice in prolactinoma without acute visual disorders.

✓ **Question 4:** Insufficiency of other hypophyseal functions including adrenal stimulation; risk of Addisonian crisis.

Case report 5

Case history

An 80-year-old man, osteoporosis with spinal fracture, otherwise in good health, impotent for years, no libido, delayed micturition, nocturia 2x

? What clinical signs of testosterone deficiency do you recognize?

Examination findings

176 cm, 75 kg, sparse secondary hair, testis volume 14 ml right, 12 ml left, hump-back, spine sensitive to percussion, muscle atrophy, prostate enlarged on palpation

Hormone analysis

		Normal range
Testosterone	7.6	12-30 nmol/l
LH	3.2	1-10 U/l
FSH	4.3	1-10 U/l

? What laboratory parameters would you measure?

? Additional tests?

Additional tests

Osteodensitometry, osteoporosis, bone density lower by 3.5 SD compared with similar age group; TRUS: BPH, 35 ml, no suspicious areas

Diagnosis/evaluation

Late-onset hypogonadism with secondary osteoporosis

? What treatment would you order?

Therapy

Transdermal testosterone, e.g. testosterone gel 1x daily

Sprightly man with late-onset hypogonadism and severe osteoporosis (spinal fractures!). Significantly reduced testosterone, therefore indicated for therapy despite benign hyperplasia of the prostate (BPH) and normal PSA. The indication for therapy is based on determination of a low serum testosterone combined with clinical signs of androgen deficiency. Regular prostate check-ups are important. In late-onset hypogonadism short-acting testosterone formulations are preferred (e.g. transdermal testosterone) as therapy can be discontinued immediately if prostate findings are suspect.

✓ **Question 1:** Osteoporosis, loss of libido and potency, sparse secondary hair, muscle atrophy
✓ **Question 2 :** Clarify osteoporosis by determining serum alkaline phosphatase and calcium (bone metabolism, hypercalcemia), renal function, ESR, blood profile and protein-electrophoresis to exclude hematological disorders.
✓ **Question 3 :** Bone density, TRUS
✓ **Question 4:** Testosterone gel or testosterone patch

Case report 6

Case history

A 48-year old man who has had surgery of the hypophysis and receives testosterone injections which his wife administers. Interval between injections is getting shorter because of lack of efficacy (currently every 12–14 days)

? What is your interpretation of the testosterone values?

Examination findings

183 cm, 75 kg, normal secondary hair, testis volume 12 ml both sides, prostate enlarged on palpation but not suspicious

Hormone analysis

		Normal range
Testosterone	25.8	12-30 nmol/l
LH	0.7	1-10 U/l
FSH	0.7	1-10 U/l

? Why are hemoglobin and hematocrit increased?

What other information do you require?

When was the last injection? 12 days ago

Biochemical findings

		Normal range
Hemoglobin	19.6 g/dl	14-18 g/dl
Hematocrit	55%	42-50 %
HDL cholesterol	28 mg/dl	35-55 mg/dl
PSA	3.3 ng/ml	< 4 ng/ml
Bone density	Normal	
TRUS	34 ml, not suspicious	

? How would you evaluate the HDL cholesterol?

How would you advise/treat the patient?

Therapy

Patient is being overdosed, as evidenced by the polyglobulism and HDL among other things; discontinue testosterone immediately, when normalized administer 250 mg testosterone enanthate i.m. every 3 weeks; alternatively, 1000 mg testosterone undecanoate every 12 weeks. Administration only by a doctor as less risk of overdosing

Patient with secondary hypogonadism requiring shorter and shorter intervals between injection possibly to improve erectile dysfunction or neurasthenic fatigue. In addition, some patients can develop a slight fixation for the injections. Testosterone overdose can be recognized by polyglobulism (hemoglobin, hematocrit) and reduced HDL. This patient must be carefully encouraged to take a break from the testosterone and then, once his parameters are normalized, be adjusted properly by a doctor. Offer to speak about erectile dysfunction.

✓ Question 1 Unusually high for therapy and time after last injection.

✓ Question 2 Hemoglobin and hematocrit are raised as a result of the testosterone overdose; this has an adverse effect on rheology and is therefore detrimental.

✓ Question 3 HDL cholesterol is reduced due to the overdose and therefore has a detrimental effect on the atherosclerotic risk.

Case report 7

Case history

A 48-year-old man with type I diabetes mellitus for 18 years, potency disorder for 5 years although strong libido

? Reason for erectile dysfunction?

Examination findings

175 cm, 68 kg, normal general medical examination, testis volume 16 ml right side, 15 ml left side, BP (RR) 155/90 mmHg, heart rate 60/min, normal secondary hair, diminished sensitivity of the feet

What tests would you order?

Biochemical findings

Following increased or pathological: creatinine 1.2 mg/dl (norm < 0.9), HbA_{1c} 7.5 % (norm < 6.0)

? What additional tests could be performed?

Hormone analysis

		Normal range
Testosterone	14.8	12-30 nmol/l
LH	4.3	1-10 U/l
FSH	2.8	1-10 U/l

Additional tests

Papaverine test positive. Rigiscan: only rare spontaneous erections, reduced rigidity

Diagnosis/evaluation

Erectile dysfunction resulting from neuropathy due to diabetes mellitus

Therapy

Phosphodiesterase-5-inhibitor, optimize diabetes treatment

Diabetic with erectile dysfunction but no testosterone deficiency. Androgen treatment not sensible in this case. Not every case of disturbed potency can be attributed to testosterone deficiency and is improved by testosterone. In this patient, symptomatic treatments other than testosterone would be more successful. In addition, the increased glycosylated hemoglobin level (HbA_{1c}) indicates that the diabetes has not been optimally controlled.

✓ Question 1 Neuropathy as a result of diabetes mellitus.

✓ Question 2 The erection and rigidity problem could be further investigated using the papaverine test and RigiScan®.

Case report 8

Case history

A 15-year-old boy, referred because puberty failed to start

? What laboratory parameters should be checked?

Examination findings

176 cm, 72 kg, leptosomatic build (slight), appears prepubescent, no facial hair, no pubic hair, testis volume 4 ml both sides, high voice, no gynecomastia, no hyposmia

? Additional tests?

Biochemical findings

		Normal range
Hemoglobin	13.6	14-18 g/dl
Alkaline phosphatase	352	< 170 U/l

Hormone analysis

Testosterone	2.5	12-30 nmol/l
LH	1.1	1-10 U/l
FSH	0.7	1-10 U/l

Additional tests

LHRH test: inadequate increase in LH and FSH, determination of bone age (X-ray of left hand): bone age 13.5 years, epiphyses still open

? Diagnosis?

Diagnosis/evaluation

Delayed puberty, differential diagnosis: hypothalamic damage, e.g. idiopathic hypogonadotropic hypogonadism (IHH)

Therapy

100 mg testosterone enanthate every 3 weeks, then 3 months without treatment to check for spontaneous onset of puberty; if necessary repeat cycle. After max. 1 year consult an endocrinologist

Boy with delayed puberty. Alkaline phosphatase is increased as the boy is still growing; this is confirmed by X-ray finding showing open epiphyses. To date we still do not know if this is a case of constitutional growth delay ('late developer') or whether the cause is organic, e.g. damage to hypothalamus or hypophysis. This will be clarified in due course. To reduce emotional stress testosterone therapy will be carried out for 3 months. This often initiates spontaneous onset of puberty. If this does not happen a specialist should be consulted. In rare cases the cause may be organic (e.g. a craniopharyngioma).

✓ Question 1 Increased alkaline phosphatase indicates that the boy is still growing.

✓ Question 2 Large increase in gonadotropins in the LHRH test indicates constitutional delayed development. Determining bone age will reveal extent of retarded development and potential for growth.

Case report 9

Case history

A 60-year-old doctor in first year of retirement after handing over practice to son, complains of neurasthenic fatigue and lack of drive, loss of libido, sleeping disorders; physical debility particularly noticeable when jogging

? How would you evaluate androgen status from the case history and findings?

Examination findings

184 cm, 79 kg, normal general medical examination, otherwise slight muscle atrophy, testis volume left side 17 ml, right side 19 ml; prostate enlarged on palpation, no lumps

Biochemical findings

		Normal range
Hemoglobin	15.1	14-18 g/dl
Hematocrit	44	42-50 %
Serum GOT	16	< 20 U/l
PSA	1.5	< 4 ng/dl

Hormone analysis

		Normal range
Testosterone	12.8	12-30 nmol/l
LH	4.6	1-10 U/l
FSH	6.2	1-10 U/l

Additional tests

TRUS: prostate 36 ml, no suspicious areas

? Would it be sensible to administer testosterone injections?

Diagnosis/evaluation

Late-onset hypogonadism
Benign hypertrophy of the prostate

Therapy

testosterone gel 1 x 5 g

Man feeling unwell, possibly due to absence of job. Biochemical findings essentially normal, testosterone in the lower normal range. No objective evidence of androgen deficiency, so limited duration of therapy offered as patient has strong desire and high hopes for treatment. Treatment is set in advance at 6 months initially, then a decision should be made whether to continue treatment after critical evaluation of its effect on symptoms.

✓ **Question 1** Presumably normal testosterone as no objective criteria for androgen deficiency.

✓ **Question 2** Success of testosterone substitution for neurasthenic fatigue and lack of drive not always predictable. However, testosterone also proven to stimulate muscle strength in older men and has beneficial effect on mood and drive. Therefore it is worth attempting.

Case report 10

Case history

A 48-year-old man, impotent for 3 years, loss of libido, tired, has been drinking three bottles of beer per evening for years, sometimes more

? How would you evaluate the situation from case history, findings and laboratory parameters?

Examination findings

170 cm, 86 kg, sparse secondary hair, testis volume 15 ml left, 16 ml right, soft consistency, no gynecomastia, no hair on abdomen, erythema of palms, spider nevi

Biochemical findings

		Normal range
γ-Glutamyl transferase	124	< 25 U/l
Serum GOT	35	< 20 U/l

? Why is testosterone low and estradiol high?

Hormone analysis

		Normal range
Testosterone	8.8	12-30 nmol/l
LH	1.2	1-10 U/l
FSH	2.8	1-10 U/l
Estradiol	130	< 85 pmol/l

? Additional tests?

Suspected diagnosis?

Additional tests

General examination for signs of alcohol-induced hepatitis, cirrhosis of the liver

? Recommended treatment?

Diagnosis/evaluation

Alcohol abuse with hypogonadism

Therapy

Evidence of link with alcohol abuse, after excluding cirrhosis possible testosterone substitution, e.g. transdermal testosterone gel; close monitoring of transaminases and liver function

Alcohol abuse with alcohol-induced hepatitis (serum GOT is increased). In patients like this estradiol is often high or increased and inhibits release of the gonadotropins LH and FSH. Hence, testosterone is low, also due to effect of alcohol on production of testosterone (alcohol inhibits enzymes of steroidogenesis). Before testosterone therapy, which should not be started if there is cirrhosis, liver function should be investigated more closely. Essentially the patient should be recommended to abstain from alcohol.

✓ **Question 1** Suspected alcohol abuse.

✓ **Question 2** Alcohol inhibits testicular enzymes therefore less testosterone is produced. In liver cirrhosis more androgens get into the adipose tissue because less are eliminated via the liver. In the adipose tissue androgens are converted to estradiol.

✓ **Question 3** General examination of liver function.

✓ **Question 4** Administer testosterone with care. If there is further alcohol abuse even testosterone will not improve libido and potency.

Case report 11

Case history

A 26-year-old man, referred because unable to have children, otherwise in own words 'fighting fit' and practices a lot of sport

? What is your suspected diagnosis according to case history, findings and semen analysis?

Examination findings

186 cm, 96 kg, very muscular and athletic, testis volume 12 ml both sides, acne on shoulders, active body-builder, denies taking anabolic steroids

Semen analysis

Sperm count	3.8	> 20 x 10^6/ml
Motility	35	> 40 %
Morphology	43	> 40 %

? What is your interpretation of the hormone analysis?

Hormone analysis

		Normal range
Testosterone	7.3	12-30 nmol/l
LH	0.2	1-10 U/l
FSH	0.2	1-10 U/l

Biochemical findings

HDL cholesterol	8	38-65 mg/dl
Hemoglobin	19.1	14-18 g/dl
Hematocrit	58	42-50 %

? How do you interpret the laboratory parameters?

Diagnosis/evaluation

Strong suspicion of anabolic steroid abuse

? What treatment would you recommend?

Therapy

Stop using anabolics, wait and check the semen again in 3 months at the earliest

This example should make you more aware of the clinical signs and biochemical evidence of anabolic abuse. Despite low testosterone levels therapy not indicated as the patient obviously not lacking in androgens. Anabolic depot preparations may still be working that have not been detected in the testosterone assay. Exogenous androgens suppress the gonadotropins and hence inhibit testicular function, including spermatogenesis.

✓ **Question 1** Rather small testes, acne, muscular body, active body-builder, oligoasthenozoospermia are consistent with anabolic administration.

✓ **Question 2** Suppression of gonadotropins by anabolics. Testosterone levels are low as the gonadotropins are low.

✓ **Question 3** HDL is low, hemoglobin and hematocrit are increased, which is typical of anabolic abuse.

✓ **Question 4** None! Confirm and wait for hypophyseal gonadotropin secretion to recover spontaneously. This may take 6-12 months.

Case report 12

Case history

A 45-year-old type I diabetic, kidney transplant 3 years ago, transplanted kidney functions well, hypertension, erectile dysfunction for a few years. Medication: prednisone, cyclosporin A, atenolol, enalapril, insulin

? What do you think caused the erectile dysfunction?

Examination findings

175 cm, 70 kg, moderate general medical examination, testes volume 14 ml right, 12 ml left, hunch-back, muscle atrophy, BP (RR) 155/90 mmHg, heart rate 60/min

Biochemical findings

The following were increased or pathological:
creatinine 1.8 mg/dl, HbA_{1c} 6.5 %

? What is your interpretation of the low testosterone levels?

Hormone analysis

		Normal range
Testosterone	9.8	12-30 nmol/l
LH	3.1	1-10 U/l
FSH	2.3	1-10 U/l

? Will testosterone improve the erectile dysfunction?

Causes of erectile dysfunction?

Neuropathy as a result of diabetes mellitus, vascular with arteriosclerosis or steal phenomenon of the kidneys, drug treatment with beta-blockers, endocrine therapy with testosterone deficiency

Additional tests

Papaverine test, if necessary RigiScan

? Would you treat with testosterone?

Diagnosis/evaluation

Erectile dysfunction caused by multiple factors, hypogonadism with general (systemic) illness

Therapy

Testosterone gel, phosphodiesterase-5-inhibitor, change beta-blocker, optimize hypertension (e.g. diuretic)

Complicated patient with kidney transplant. As general impotence considerably reduces quality of life so remedial action should be taken. Obviously multiple reasons for erectile dysfunction: (1) There are very many potential causes of impotence, some of which cannot be changed (e.g. steal phenomenon). (2) Because of multifactorial genesis treatment should be on several levels (hormone therapy plus phosphodiesterase-5-inhibitor); (3) Testosterone is not usually contraindicated in kidney transplant. It has a beneficial effect on erythropoiesis, musculature, physical ability, libido and potency; (4) Drugs may have an adverse effect on potency, especially beta-blockers. However, with complicated patients treatment should only be changed in consultation with the general practitioner.

✓ **Question 1** Hypertension, diabetes mellitus, renal insufficiency, medication (e.g. corticosteroids, beta-blockers), arteriosclerosis, kidney transplant.

✓ **Question 2** General disease, slight renal insufficiency, drugs (prednisone).

✓ **Question 3** Hardly, as caused by many factors.

✓ **Question 4** Testosterone is indicated for the testosterone deficiency. This is most likely to have a beneficial effect.

Case report 13

Case history
A 28-year-old man who is unable to have children

Examination findings
192 cm, 89 kg, sparse secondary hair, shaves every 2 days, testis volume 13 ml both sides, no gynecomastia

Semen analysis

Sperm count	0.8	> 20 x 10^6/ml
Motility	15	> 40 %
Morphology	22	> 40 %

Hormone analysis

		Normal range
Testosterone	26.3	12-30 nmol/l
LH	17.4	1-10 U/l
FSH	12.1	1-10 U/l

Suspected diagnosis?

Additional tests
DHT: 4.5 nmol/l (slightly increased)
Estradiol: 125 pmol/l (slightly increased)
SHBG: 55 nmol/l (borderline high)
Gene analysis for mutation in androgen receptor gene

Diagnosis/evaluation
Mild androgen resistance (Reifenstein's syndrome)

Therapy
Attempt to achieve better androgenization with high-dose testosterone therapy, 250 mg i.m. testosterone depot every 2 weeks, recommend assisted fertilization (e.g. intracytoplasmic sperm injection), adoption

? How do you explain the discrepancy between clinical signs of androgen deficiency and high testosterone levels?

? Additional tests?

Patient with a very rare disease. The clinical symptoms (sparse secondary hair, gynecomastia, severe oligoasthenoteratozoospermia) indicate androgen deficiency. Androgens are surprisingly high with simultaneously increased gonadotropins. A finding such as this should lead one to consider androgen resistance if confirmed by a second blood sample. This patient is suffering from mild androgen resistance as he has male genitalia. Severe androgen resistance leads to intersexual genitals or complete feminization which would have been picked up when the patient was born or when referred to a gynecologist. It is possible nowadays to carry out gene diagnosis in specialist clinics. Better virilization can be obtained in some patients with high-dose testosterone therapy.

✓ **Question 1** This combination of symptoms should always lead one to consider androgen resistance.

✓ **Question 2** To strengthen the suspected diagnosis other steroids such as SHBG can be measured. Diagnosis can only be confirmed in specialist clinics using gene technology.

Case report 14

Case history

A 53-year-old dialysis patient with interstitial nephritis caused by abuse of analgesics, hypertension. Impotent for years
Medication: metoprolol, captopril, prazosin, vitamin D_3 preparation, calcium carbonate

? What clinical symptoms could be improved by androgen therapy?

Examination findings

175 cm, 70 kg, moderate general medical examination, testis volume 14 ml right, 12 ml left, hunch-back, muscle atrophy, BP (RR) 160/95 mmHg, heart rate 60/min, small prostate

Biochemical findings

The following were increased or pathological: creatinine 3.8 mg/dl, urea 180 mg/dl, hemoglobin 10.1 g/dl, hematocrit 34 %, ESR 38 mm

Hormone analysis

		Normal range
Testosterone	8.9	12-30 nmol/l
LH	3.1	1-10 U/l
FSH	2.3	1-10 U/l

Additional tests

Papaverine test, if necessary RigiScan

? Why are testosterone depot injections not suitable?

Diagnosis/evaluation

Erectile dysfunction due to a number of factors, hypogonadism with general illness

Therapy

Testosterone substitution with testosterone gel, phosphodiesterase-5-inhibitor

Dialysis patient on a number of medications, renal anemia and androgen deficiency as a result of chronic renal insufficiency. Testosterone medication has a beneficial effect on renal osteopathy, muscle atrophy, anemia and potency. It is not contraindicated.

✓ **Question 1** Osteoporosis and muscle atrophy, quality of life.

✓ **Question 2** Intramuscular injections are unsuitable as the patient is fully heparinized three times a week during dialysis and intramuscular injections have a high risk of causing large hematoma.

Case report 15

Case history

A 65-year-old patient who underwent prostatectomy 6 years ago for carcinoma of the prostate (T1 N0 M0). As tumor local no further therapy required; since then no recurrences. Now loss of potency and libido, muscles weak; GP detected osteoporosis

Examination findings

173 cm, 75 kg, normal general medical examination, testis volume 13 ml right, 14 ml left, muscle atrophy, BP (RR) 160/85 mmHg, heart rate 72/min

Biochemical findings

Normal, PSA not detected

Hormone analysis

		Normal range
Testosterone	6.9	12-30 nmol/l
LH	8.1	1-10 U/l
FSH	7.7	1-10 U/l

Additional tests

TRUS: prostate absent, no evidence of local recurrence or lymph nodes

Diagnosis/evaluation

Late-onset hypogonadism, prostate resection for carcinoma (T1 N0 M0)

Can androgens be administered?

Therapy

If decide on testosterone substitution, then short-acting preparations, e.g. testosterone gel

? Why is testosterone gel particularly suitable in this situation?

Complex situation for which there is no correct answer owing to lack of studies. The pros and cons will have to be carefully weighed up with the patient. Depending on the psychological state of the patient (e.g. severe, painful osteoporosis or potency disorder) androgen substitution can be started. Then PSA must be monitored closely. Since PSA is not detected, and there has been no recurrence for 6 years, one can conclude that the prostatectomy was complete and has now healed.

✓ **Question 1** Substitution with testosterone gel can be discontinued at any time, unlike injections with testosterone esters.

Case report 16

Case history

A 14-year-old boy, referred for hormone therapy for excessive growth; mother is 188 cm, father is 190 cm

? What does the alkaline phosphatase tell you?

Examination findings

191 cm, 74 kg, leptosomatic physique, prepubescent, no facial hair, no pubic hair, testis volume 4 ml both sides, high voice, no gynecomastia

Biochemical findings

		Normal range
Alkaline phosphatase	352	< 170 U/l

Hormone analysis

		Normal range
Testosterone	3.2	12-30 nmol/l
LH	1.1	1-10 U/l
FSH	1.5	1-10 U/l
IGF-1	620	250-760 ng/dl

? Is the serum testosterone normal?

Additional tests

LHRH test: increase in LH and FSH appropriate for age; oral glucose tolerance test with growth hormone determination: GH < 0.2 ng/dl (normal); bone age determination (X-ray left hand): bone age 13.5 years, epiphyses still open

What treatment would you suggest?

? Why do you need to determine bone age?

Diagnosis/evaluation

Excessive growth

Therapy

500 mg testosterone enanthate every week for 3-4 months, monitor epiphyses

? When should you introduce therapy?

The boy has constitutional excessive growth and parents who are also excessively tall. The presumed final height will be over 2 m (the rule for boys is mean of parents' heights plus 10 %). After excluding other causes of excessive growth, high-dose testosterone therapy may reduce final height. Increased secretion of growth hormone (GH) as in gigantism can be excluded with the oral glucose tolerance test (oGTT) and IGF-I. High-dose testosterone therapy for reducing final height should only be administered by a specialist in endocrinology. It is important to start therapy early.

✓ **Question 1** The boy is still growing.
✓ **Question 2** Adequate for a 14-year-old boy.
✓ **Question 3** Estimate potential for growth based on epiphyses, determine biological age.
✓ **Question 4** If therapy to be given, this should start immediately, otherwise there will be no significant reduction in final height.

Case report 17

Case history

A 55-year-old man with sclerodermatitis, long-term corticosteroid treatment, currently on 35 mg prednisone daily, no potency disorders; GP found bone density to be normal. Referred to urologist for preventive prostate screening

? Why is LH low?

Examination findings

173 cm, 75 kg, Raynaud's syndrome, calcinosis cutis, muscle atrophy, prostate normal on palpation

Which hormones would you have analyzed?

Hormone analysis

		Normal range
Testosterone	10.4	12-30 nmol/l
LH	1.1	1-10 U/l
FSH	2.7	1-10 U/l

Diagnosis/evaluation

Sclerodermatitis

Secondary hypogonadism as a result of corticosteroid therapy

? Why would it be a good idea to give testosterone therapy?

Therapy

Substitution indicated (prophylaxis of muscle atrophy and osteoporosis), 250 mg testosterone enanthate i.m. every 3 weeks, alternatively 1000 mg testosterone undecanoate every 12 weeks or 160-200 mg/day oral testosterone undecanoate; transdermal administration of testosterone is less suitable because the skin is sensitive (skin irritation!).

Men on long-term corticosteroid therapy often have low testosterone levels. On higher doses corticosteroids react with the androgen receptor and inhibit LH secretion. There is a simultaneous fall in testicular production of testosterone. Since corticosteroids have a catabolic effect, muscle atrophy and osteoporosis often develop and this effect is intensified by the androgen deficiency. Neither corticosteroid therapy nor the underlying disease are contraindications of testosterone therapy. The formulation can be changed, depending on the underlying disease and the treatment. Since this patient has sclerodermatitis and the skin is inflamed, intramuscular administration is recommended. In systemic lupus erythematosus with thrombopenia and tendency to develop hematoma, but where the skin is not inflamed, transdermal treatment is preferred.

✓ Question 1 Negative feedback of corticosteroids to hypophysis.

✓ Question 2 Muscle atrophy, prophylaxis of osteoporosis

Case report 18

Case history

A 33-year-old man with orchiectomy left side 2 years ago due to Leydig cell tumor. When he was unable to have children his azoospermia led to a biopsy of the testes, which revealed Leydig cell tumors in both testes. Apparently no hormone tests were carried out. Patient is very unsure and wants a 'second opinion'

? What findings lead you to suspect Klinefelter's syndrome?

Examination findings

183 cm, 75 kg, testis right side 3 ml, no larger before biopsy; slight gynecomastia, small prostate

? Was it sensible to perform a biopsy of the testis?

Hormone analysis

		Normal range
Testosterone	4.4	12-30 nmol/l
LH	18.1	1-10 U/l
FSH	25.7	1-10 U/l

What is your suspected diagnosis and how would you prove it?

? Additional tests?

Diagnosis/evaluation

Chromosome analysis: 47,XXY, Klinefelter's syndrome

Does the patient have a Leydig cell tumor?

No! Klinefelter patients suffer from relative hyperplasia of the Leydig cells

Would you give testosterone substitution?

Therapy

Substitution indicated (prophylaxis of muscle atrophy and osteoporosis), 250 mg testosterone enanthate i.m. every 3 weeks

Patient who has had endocrinological consultation. Prior treatment incorrect. Because he had azoospermia a biopsy was ordered on the testes which were approximately 3 ml in size without doing a hormone analysis in advance. In the absence of a suspected diagnosis from the urologist, the less-experienced pathologist obviously confused the testis histology and concluded that this was a case of Leydig cell tumor. The combination of azoospermia and very small testes should always lead one to consider Klinefelter's syndrome and a hormone and chromosome analysis should be carried out. The increased FSH would have made the biopsy superfluous. Klinefelter patients often have relative hyperplasia of the Leydig cells since the seminiferous tubules are very small and contain hardly any gametes, which is why the testes are also very small. The inexperienced pathologist mistook the Klinefelter's for a tumor of the Leydig cells. This patient could have been spared the unilateral orchiectomy and the feeling he was suffering from 'cancer'. Life-long testosterone substitution should be given.

✓ Question 1 Small testes, azoospermia, gynecomastia.
✓ Question 2 Definitely not!
✓ Question 3 Osteodensitometry.

Case report 19

Case history

A 23-year-old man, delayed puberty, testosterone therapy from 15 to 20 years of age, then discontinued by GP as sufficiently virilized; now loss of libido and potency

Examination findings

186 cm, 76 kg, unambiguously male disposition, sparse secondary hair, penis 6 cm in length when extended,
small scrotum, testes not palpable either side, prostate hardly palpable

? Suspected diagnosis?

Hormone analysis

		Normal range
Testosterone	0.8	12-30 nmol/l
LH	33.4	1-10 U/l
FSH	41.7	1-10 U/l

? How would you interpret the result?

Suspected diagnosis? How would you proceed?

Cryptorchidism vs. anorchidism

hCG test:

5000 IU hCG i.m., testosterone beforehand and 72 h afterwards; norm: 1.5-2.5-fold increase

Result: testosterone basal 0.8 nmol/l

after 72 h 1.0 nmol/l

Diagnosis/evaluation

Congenital anorchidism

Therapy

Life-long substitution with testosterone, e.g. 250 mg testosterone enanthate i.m. every 3 weeks or 1000 mg testosterone undecanoate i.m. every 12 weeks

How do you proceed if the hCG test is positive?

Delayed onset of puberty is a symptom, not a diagnosis. In the majority of cases it is caused by constitutional growth delay ('late developer') which is normally harmless, but it can also be caused by organic diseases with hypogonadism which manifests when puberty fails to set in. Hence, a complete diagnosis should be made after a short phase of treatment; if necessary this may include MRI of the hypophysis. Obviously, no one had examined this patient's testicles before, otherwise they would have noticed they were missing. It is important to differentiate between cryptorchidism and anorchidism. As the hCG test is negative one can assume that this is a case of anorchidism. A positive hCG test indicates that there is testicular tissue still present in the pelvis or inguinal canal which has to be located. Undescended testes have a high tendency to degenerate, and are therefore removed, especially since they do not function.

✓ **Question 1** The clinical finding of absent testes indicates cryptorchidism or anorchidism.

✓ **Question 2** The negative hCG test confirms anorchidism.

Case report 20

Case history

A 68-year-old man with neurasthenic vitality deficiency, tiredness, decreasing potency and loss of libido

? What clinical signs of androgen deficiency do you recognize?

Examination findings

176 cm, 78 kg, normal secondary hair growth, slight muscle atrophy, testis volume 15 ml both sides, prostate slightly enlarged on palpation but not suspicious

What tests would you order?

Hormone analysis

		Normal range
Testosterone	7.8	12-30 nmol/l
LH	3.7	1-10 U/l
FSH	5.7	1-10 U/l

? What laboratory parameters are you interested in?

Biochemical findings

		Normal range
HDL cholesterol	46	38-65 mg/dl
Hemoglobin	14.1	14-18 g/dl
Hematocrit	43	42-50 %
PSA	2.7	< 4 ng/dl

? What additional tests would you recommend?

Additional tests

TRUS: BPH, 28 ml

Osteodensitometry: osteopenia, bone density lower than equivalent age-group by 1.8 SD

? What treatment would you give?

Diagnosis/evaluation

Late-onset hypogonadism

Therapy

Androgen substitution, e.g. transdermal with testosterone gel

Typical patient with late-onset hypogonadism. He shows several signs of androgen deficiency. The decision to treat is therefore easy, especially since there are no recognizable contraindications.

✓ Question 1 Loss of potency and libido, neurasthenic fatigue, tiredness, muscle atrophy.

✓ Question 2 If there was a severe metabolic disorder with low HDL and high LDL, androgen therapy could be considered. The blood profile is interesting because of potential androgen-deficiency anemia or polyglobulism. PSA is important to evaluate the prostate.

✓ Question 3 TRUS, bone density.

✓ Question 4 The advantage of testosterone gel is that therapy can be discontinued immediately. Testosterone injections are another possibility.

Case report 21

Case history

A 38-year-old man with persistent growth, bone pain, osteoporosis and open epiphyses; unable to have children; potency satisfactory but libido only moderate, sufficiently virilized

Examination findings

190 cm, 97 kg, eunuchoid proportions, normal secondary hair growth, testis volume 8 ml both sides, no gynecomastia

Hormone analysis

		Normal range
Testosterone	18.1	12-30 nmol/l
LH	5.6	1-10 U/l
FSH	17.1	1-10 U/l
Estradiol	< 10	25-85 pmol/l

Semen analysis

Severe oligoasthenozoospermia

Estrogen deficiency in a man?

Additional tests

Refer to a specialist clinic with query of aromatase deficiency (mutation in aromatase gene)

Therapy when diagnosis confirmed

Therapy

Estrogen substitution, e.g. 0.5-1 x 2-mg tablet estradiol

Could estriol be substituted?

No! Does not work in bones

Extremely rare case report from literature of patient with aromatase deficiency. Difficult to diagnose as estradiol is not routinely determined (how many superfluous estradiol determinations would have to be carried out to ascertain one significant finding?). The osteoporosis in this patient started very early and there was no obvious cause, which is why rare diseases have to be considered, including estrogen deficiency or androgen receptor defect. Gonadotropins are also relatively high with high testosterone which already indicates lack of estradiol. The patient must be referred to a specialist clinic for gene diagnosis. As far as treatment is concerned, administration of estradiol is indicated, estriol is not suitable as it does not work in bones.

Case report 22

Case history

A 57-year-old man on testosterone substitution for bilateral orchiectomy 15 years ago following an accident. Complains of severe fluctuations in mood, physical ability and libido from testosterone injections every 18-20 days

Examination findings

175 cm, 79 kg, normal general medical examination, normal secondary hair, prosthetic testicles both sides, prostate normal

Hormone analysis (14 days after injection)

		Normal range
Testosterone	7.8	12-30 nmol/l
LH	15.3	1-10 U/l
FSH	22.5	1-10 U/l

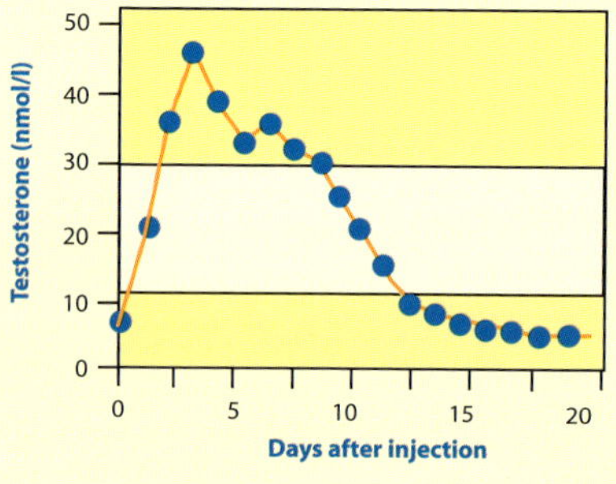

Clinico-chemical parameters

All normal

What other tests are still required?

Osteodensitometry
TRUS

What treatment would you recommend?

Testosterone gel, alternatively 1000 mg testosterone undecanoate every 12 weeks

Patient with clear indication for testosterone substitution. Previously correctly treated with testosterone enanthate but finds the fluctuating levels very unpleasant. The blood sample shows this (see Figure). This also reflects the typical kinetics of the preparation which sensitive patients experience as fluctuations in mood, performance and sexual activity. If this patient is changed to transdermal therapy with testosterone gel, the serum testosterone levels should balance out, leading to stable functional capacity.

Case report 23

Case history

A 23-year-old man requesting the 'male pill'

What do you do?

In principle it is possible to give men hormonal contraceptives.
This procedure has been used in studies:
200 mg testosterone enanthate i.m. per week sometimes with additional gestagen administration
Success rate: > 75 % azoospermia after 3 months
Problem:
(1) The long-term effects of a testosterone overdose are still unknown
Prostate growth?
Adverse effect on lipid metabolism?
(2) Not a safe prophylactic for everyday use

Procedure

Counsel and inform the patient
Indicate alternatives
Cannot prescribe such treatment since:
(1) Internationally no preparation has been licensed for this indication
(2) It is not clear who is financially liable if treatment fails

? How do hormone contraceptives work in men?

? How safe is the procedure?

? What is the current legal situation?

Subject 'Male pill'. The doctor should be aware that although high-dose testosterone therapy can be used in men as hormonal contraception, it is not safe, there is no long-term experience available and in terms of safety and liability there are many gray areas. Therefore, the doctor can advise the patient and decide for himself whether he is prepared to accept responsibility for this form of contraception.

✓ **Question 1:** High-dose testosterone suppresses LH and FSH, so spermatogenesis is not stimulated and intratesticular testosterone levels fall. Sperm production ceases resulting in azoospermia. Symptoms of androgen deficiency can be prevented by exogenously administered testosterone. Gestagens can intensify suppression of LH and FSH.

✓ **Question 2:** According to studies in Asian men, approximately as safe as the "female pill".

✓ **Question 3:** Not at all clear.

Case report 24

Case history

A 47-year-old man with increasing loss of drive, general lack of enthusiasm, subjective feeling of unrest, frequently exhausted. Problem made worse by having to cope with professional demands

Examination findings

177 cm, 74 kg, muscular, good physical shape, normal secondary hair growth, testis volume 14 ml both sides, prostate normal

Biochemical findings

		Normal range
ESR	9	< 20 mm/h
Hemoglobin	14.2	14-18 g/dl
PSA	0.7	< 4 ng/dl

Hormone analysis

		Normal range
Testosterone	14.1	12-30 nmol/l
LH	2.8	1-10 U/l
FSH	3.2	1-10 U/l

Diagnosis/evaluation

Feels unwell but no testosterone deficiency

Therapy

Discuss circumstances, if strong desire for treatment and to pave the way testosterone gel treatment can be attempted for a limited period only until symptoms improve

This patient's main symptoms are neurasthenic fatigue and lack of drive. However, there are no objective signs of androgen deficiency and testosterone levels are not low. A psychological evaluation would be sensible as the patient may also be suffering from depression. Many patients are initially loathe to undergo a psychological assessment. Prescribing testosterone for a defined period can help pave the way to further measures.

Literature

8. Literature

1. Aakvaag A, Vogt JH: Plasma testosterone values in different forms of testosterone treatment. Acta Endocrinol. 60: 537-42, 1969

2. Aarsen RSR: Micropenis: medical and surgical implications. J.Urol. 152: 4-14, 1994

3. Adami S, Fossaluzza V, Rossini M, et al: The prevention of corticosteroid-induced osteoporosis with nandrolone decanoate. Bone Miner. 15: 72-81, 1991

4. Adler RA: Clinically important effects of alcohol on endocrine function. J.Clin.Endocrinol.Metab. 74: 957-60, 1992

5. Ahmed SR, Boucher AE, Manni A, Santen RJ, Bartholomew M, Demers LM: Transdermal testosterone therapy in the treatment of male hypogonadism. J.Clin.Endocrinol.Metab. 66: 546-51, 1988

6. Aiman J, Griffin JE: The frequency of androgen receptor deficiency in infertile men. J.Clin.Endocrinol.Metab. 54: 725-32, 1982

7. Alexander GM, Swerdloff RS, Wang C, Davidson T, McDonald V, Steiner B, Hines M: Androgen-behavior correlations in hypogonadal men and eugonadal men. I. Mood and response to auditory stimuli. Horm.Behavior 31: 110-9, 1997

8. Alexander GM, Swerdloff RS, Wang C, Davidson T, McDonald V, Steiner B, Hines M: Androgen-behavior correlations in hypogonadal men and eugonadal men. II. Cognitive abilities. Horm.Behavior 33: 85-94, 1998

9. Alexander GM: Androgens and cognitive function. In: Bhasin S, Gabelnick HL, Spieler JM, Swerdloff RS, Wang C, Kelly C: Pharmacology, Biology, and Clinical Applications of Androgens. John Wiley, New York, 169-77, 1996

10. Ammus SS: The role of androgens in the treatment of hematologic disorders. Adv.Intern.Med. 34: 191-208, 1989

11. Amrhein JA, Klingensmith GJ, Walsh PC, McKusick VA, Migeon CJ: Partial androgen insensitivity. The Reifenstein syndrome revisited. N.Engl.J.Med. 297: 350-6, 1977

12. Anderson DC: Osteoporosis in men. Br.Med.J. 305: 489-90, 1992

13. Anderson FH, Francis RM, Peaston RT, Wastell HJ: Androgen supplementation in eugonadal men with osteoporosis: effects of six months' treatment on markers of bone formation and resorption. J.Bone Mineral Res. 12: 472-8, 1997

14. Anderson RA, Baird DT: Male contraception. Endocr.Rev. 23: 735-62, 2002

15. Anderson RA, Bancroft J, Wu FCW: The effects of exogenous testosterone on sexuality and mood of normal men. J.Clin.Endocrinol.Metab. 75: 1503-7, 1992

16. Anderson RA, Martin CW, Kung AWC, Everington D, Pun TC, Tan KCB, Bancroft J, Sundaram K, Moo-Young AJ, Baird DT: 7α-Methyl-19-Nortestosterone maintains sexual behavior and mood in hypogonadal men. J.Clin.Endocrinol.Metab. 84: 3556-62, 1999

17. Aversa A, Isidori AM, Spera G, Lenzi A, Fabbri A: Androgens improve cavernous vasodilation and response to sildenafil in patients with erectile dysfunction. Clin.Endocrinol. 58: 632-638, 2003

18. Babar A, Fares HN, Plakogiannis FM: In vitro release of testosterone from suppository bases and in vivo absorption studies in human males. J.Pharmaceut.Sci. 83: 389-91, 1993

19. Bagatell CJ, Bremner WJ: Androgens in men - uses and abuses. N.Engl.J.Med. 334: 707-14, 1996

20. Baisley KJ, Boyce MJ, Bukofzeri 5, Pradhani R, Warrington SJ: Pharmacokinetics, safety and tolerabihity of three dosage regimens of buccal adhesive testosterone tablets in healthy men suppressed with leuprorelin. J.Endocrinol. 175: 813-819, 2002

21. Balkau B, Charles MA. Comment on the provisional report from the WHO consultation. European Group for the Study of Insulin Resistance (EGIR). Diabet Med 16: 442-443, 1999

22. Bals-Pratsch M, Knuth UA, Yoon Y-D, Nieschlag E: Transdermal testosterone substitution therapy for male hypogonadism. Lancet 4: 943-6, 1986

23. Barrett-Connor E, Khaw KT, Yen SS. Endogenous sex hormone levels in older adult men with diabetes mellitus. Am J Epidemiol. 132:895-901, 1990

24. Barrett-Connor E, Khaw KT: Endogenous sex hormones and cardiovascular disease in men. Circulation 78: 539-45, 1988

25. Basaria S, Müller DC, Carducci MA, Egan J, Dobs AS: Hyperglycemia and insulin resistance in men with prostate carcinoma who recieve androgen deprivation therapy. Cancer 106: 581 – 8, 2006

26. Bauer J, Koch W: Über die therapeutische Wirksamkeit des reinen Hodenhormons. Wien.Med.Wschr. 87: 592-4, 1937

27. Baumgartner RN, Waters DL, Gallagher D, Morley JE, Garry PJ: Predictors of skeletal muscle mass in elderly men and women. Mech.Age.Develop. 107: 123-36, 1999

28. Becker KL, Hoffman DL, Albert A, Underdahl LO, Mason HL: Klinefelter's syndrome. Arch.Int.Med. 118: 314-21, 1966

29. Behre HJM, Bohmeyer J, Nieschlag E: Prostate volume in testosterone-treated and untreated hypogonadal men in comparison to age-matched normal controls. Clin.Endocrinol. 40: 341-9, 1994

30. Behre HJM, Kliesch S, Leifke E, Link TM, Nieschlag E: Long-term effect of testosterone therapy on bone mineral density in hypogonadal men. J.Clin.Endocrinol.Metab. 82: 2386-90, 1997

31. Behre HJM, Nieschlag E: Testosterone buciclate (20 Aet-1) in hypogonadal men: pharmacokinetics and pharmacodynamics of the new long-acting androgen ester. J.Clin.Endocrinol.Metab. 75: 1204-10, 1992

32. Behre HM, Bohmeyer J, Nieschlag E: Prostate volume in testosterone-treated and untreated hypogonadal men in comparison to age-matched normal controls. Clin.Endocrinol. 40: 341-9, 1994

33. Belkien LD, Schürmeyer T, Hano R, Gunnarsson PO, Nieschlag E: Pharmacokinetics of 19-nortestosterone esters in normal men. J.Steroid Biochem. 22: 623-9, 1985

34. Belville C, Josso N, Picard J-Y: Persistence of Müllerian derivatives in males. Am.J.Med.Gen. 89: 218-23, 199

35. Beranova M, Oliveira LMB, Bedecarrats GY, Schipani E, Vallejo M, Ammini AC, Quintos JB, Hall JE, Martin KA, Hayes FJ, Pitteloud N, Kaiser UB, Crowley WF, Seminara SB: Prevalence, phenotypic spectrum, and modes of inheritance of gonadotropin-releasing hormone receptor mutations in idiopathic hypogonadotropic hypogonadism. J.Clin.Endocrinol.Metab. 86: 1580-6, 2001

36. Bernasconi S, Ghizzoni L, Panza C, Volta C, Caselli G: Congenital anorchia: natural history and treatment. Horm.Res. 37 (Suppl.): 50-4, 1992

37. Berthold AA: Transplantation der Hoden. Arch.Anat.Physiol.Wiss.Med. 42-6, 1849

38. Bevan JS, Webster J, Burke CW, Scanlon MF: Dopamine agonists and pituitary tumor shrinkage. Endocr.Rev. 13: 220-40, 1992

39. Bhasin S, Bagatell CJ, Bremner WJ, Plymate SR, Tenover JS, Korenman SG, Nieschlag E: Issues in testosterone replacement in older men. J.Clin.Endocrinol.Metab. 83: 3435-3448, 1998

40. Bhasin S, Buckwalter JG: Testosterone supplementation in older men: a rationale idea whose time has not yet come. J.Androl. 22: 718-31, 2001

41. Bhasin S, Storer T, Berman N, et al: The effects of supraphysiologic doses of testosterone on muscle size and strength in normal men. N.Engl.J.Med. 335: 1-7, 1996

42. Bhasin S, Storer TW, Berman N, et al: Testosterone replacement increases fat-free mass and muscle size in hypogonadal men. J.Clin.Endocrinol.Metab. 82: 407-13, 1997

43. Bhasin S, Swerdloff RS, Steiner B, et al: A biodegradable testosterone microcapsule formulation provides uniform eugonadal levels of testosterone for 10 - 11 weeks in hypogonadal men. J.Clin.Endocrinol.Metab. 74: 75-83, 1992

44. Bhasin S, Woodhouse L, Casaburi R, Singh AB, Bhasin D, Berman N, Chen X, Yarasheski KE, Magliano L, Dzekov C, Dzekov J, Bross R, Phillips J, Sinha-Hikim I, Shen R, Storer TW. Testosterone dose-response relationships in healthy young men. Am J Physiol Endocrinol Metab. 281:E1172-81, 2001

45. Bhasin S, Woodhouse L, Storer TW: Proof of the effect of testosterone on skeletal muscle. J.Endocrinol. 170: 27-38, 2001

46. Bilezikian JP: Osteoporosis in men. J.Clin.Endocrinol.Metab. 84: 3431-4, 1999

47. Biro FM, Lucky AW, Huster GA, Morrison JA: Pubertal staging in boys. J.Pediatr. 127: 100-2, 1995

48. Björntorp P: Abdominal obesity and the development of noninsulin-dependent diabetes mellitus. Diabetes/Metabolism Reviews 4: 615-22, 1988

49. Block RI, Farinpour R, Schlechte JA: Effects of chronic marijuana use on testosterone, luteinizing hormone, follicle-stimulating hormone, prolactin and cortisol in men and women. Drug Alcohol Dependence 28: 121-8, 1992

50. Bluher M, Stumvoll M. Metabolic syndrome - myths, mechanisms, management. Dtsch Med Wochenschr. 131:1167-72, 2006

51. Blumenthal RS, Heldman AW, Brinker JA: Acute effects of conjugated estrogens on coronary blood flow response to acetylcholin in men. Am.J.Cardiol. 80: 1021-4, 1997

52. Boland MC: The role of sex steroids in prostate carcinogenesis. J.Natl.Cancer Inst. 27: 39-66, 2000

53. Bouchard P, Wright F, Portois MC, Couzinet B, Schaison G, Mowszowicz I: Androgen insensitivity in oligospermic men: a reappraisal. J.Clin.Endocrinol.Metab. 63: 1242-6, 1986

54. Bourguignon JP: Linear growth as a function of age at onset of puberty and sex steroid dosage: therapeutic implications. Endocr.Rev. 9: 467-89, 1988

55. Boyanov MA, Boneva Z, Christov VG: Testosterone supplementation in men with type 2 diabetes, visceral obesity and partial androgen deficiency. Aging Male 6: 1-7, 2003

56. Brämswig JH, von Lengerke HJ, Schmidt H, Schellong G: The results of short-term (6 months) high-dose testosterone treatment on bone age and adult height in boys of excessively tall stature. Eur.J.Pediatr. 148: 104-6, 1988

57. Bremner WJ, Vitiello MV, Prinz PN: Loss of circadian rhythmicity in blood testosterone levels with aging in normal men. J.Clin.Endocr. 56: 1278-81, 1983

58. Breuer B, Trungold S, Martucci C, Wallenstein S, Likourezos A, Libow LS, Zumoff B: Relationsship of sex hormone levels to dependencs in activities of daily living in the frail elderly. Maturitas 39: 147-59, 2001

59. Brinkman AO, Jenster G, Kuiper GGJM, et al: Structure and function of the human androgen receptor. In: Nieschlag E, Habenicht UF: Spermatogenesis, Fertilization, Contraception. Springer Verlag, Berlin Heidelberg, 97-122, 1992

60. Brodsky IG, Balagopal P, Nair SK: Effects of testosterone replacement on muscle mass and muscle protein synthesis in hypogonadal men - a clinical research center study. J.Clin.Endocrinol.Metab. 81: 3469-75, 1996

61. Bross R, Javanbahkit M, Bhasin S: Anabolic interventions for aging-associated sarcopenia. J.Clin.Endocrinol.Metab. 84: 3420-30, 1999

62. Brown DC, Butler GE, Kelnar CJH, Wu FCW: A double blind, placebo controlled study of the effects of low dose testosterone undecanoate on the growth of small for age, prepubertal boys. Arch.Dis.Child. 73: 131-5, 1995

63. Brown-Sequard CE: Des effets des produits chez l'homme par des injections souscutanées d'un liquide retiré des testicules frais de cobaye et de chien. CR.Seamc.Soc.Biol. 1: 420-31, 1889

64. Bruch HR, Wolf L, Budde R, Romalo G, Schweikert HU: Androstenedione metabolism in cultured human osteoblast-like cells. J.Clin.Endocrinol.Metab. 75: 101-5, 1992

65. Bullmann C, Jockenhövel F: Gynäkomastie beim Mann. Fortschr.Med. 116: 18-22, 1998

66. Burris AS, Banks SM, Carter CS, Davidson DM, Sherins RJ: A long-term, prospective study of the physiologic and behavioral effects of hormone replacement in untreated hypogonadal men. J.Androl. 13: 297-304, 1992

67. Butenandt A, Hanisch G: Über Testosteron. Umwandlung des Dehydroandrosterons in Androstendiol und Testosteron; ein Weg zur Darstellung des Testosterons aus Cholesterin. Hoppe-Seyler's Z.Physiol.Chemie 237: 714-23, 1935

68. Butler GE, Sellar RE, Walker RF, Hendry M, Kelnar CJH, Wu FCW: Oral testosterone undecanoate in the management of delayed puberty in boys: pharmacokinetics and effects on sexual maturation and growth. J.Clin.Endocrinol.Metab. 75: 37-44, 1992

69. Capel B: Sex in the 90s: SRY and the switch to the male pathway. Ann.Rev.Physiol. 60: 497-523, 1998

70. Carani C, Granata ARM, Bancroft J, Marrama P: The effects of testosterone replacement on nocturnal penile tumescence and rigidity and erectile response to visual erotic stimuli in hypogonadal men. Psychoneuroendocr. 20: 743-53, 1995

71. Carani C, Qin K, Simoni M, et al: Effect of testosterone and estradiol in a man with aromatase deficiency. N.Engl.J.Med. 337: 91-5, 1997

72. Carani C, Rochira R, Faustini-Fustini M, Balestieri A, Granata ARM: Role of oestrogen in male sexual behaviour: insights from the natural model of aromatase deficiency. Clin.Endocrinol. 51: 517-24, 1999

73. Carani C, Zini D, Baldini A, Della Casa L, Ghizzoni L, Marrama P: Effects of androgen treatment in impotent men with normal and low levels of free testosterone. Arch.Sex.Behavior 19: 223-34, 1990

74. Caron P, Bennet A, Camare R, Louvet JP, Boneu JP, Sie P: Plasminogen activator inhibitor in plasma is related to testosterone in men. Metabolism 38: 1010-5, 1989

75. Chapman RM: Gonadal injury resulting from chemotherapy. Am.J.Ind.Med. 4: 149-61, 1983

76. Charny CW, Gordon JA: Testosterone rebound therapy: a neglected modality. Fertil.Steril. 29: 64-8, 1978

77. Chen Z, Maricic M, Nguyen P, Ahmann FR, Bruhn R, Dalkin BL. Cancer. Low bone density and high percentage of body fat among men who were treated with androgen deprivation therapy for prostate carcinoma. Cancer 95:2136-44, 2002

78. Cherrier MM, Asthana S, Plymate SR, Baker L, Matsumoto AM, Peskind E, Raskind M, Brodkin A, Bremner WB, Petrova A, LaTrendresse S, Craft S: Testosterone supplementation improves spatial and verbal memory in healthy older men. Neurology 57: 80-8, 2001

79. Christiansen K: Behavioural correlates of testosterone. In: Nieschlag E, Behre HM: Testosterone - Action, Deficiency, Substitution. Springer Verlag, Berlin, 107-42, 1998

80. Clark RV: History and physical examination. Endocrinol.Metab.Clin.North Am. 23: 699-708, 1994

81. Collins P, Rosano GMC, Sarrel PM: 17-beta estradiol attenuates acetylcholine-induced coronary arterial vasoconstriction in women but not in men with coronary heart disease. Circulation 92: 24-30, 1995

82. Colvard DS, Eriksen EF, Keeting PE, et al: Identification of androgen receptors in normal human osteoblast-like cells. Proc.Natl.Acad.Sci.U.S.A. 86: 854-7, 1989

83. Coronary Drug Project Research Group: Initial findings leading to modifications of its research protocol. J.Am.Med.Ass. 214: 1303-13, 1970

84. Coronary Drug Project Research Group: The coronary drug project: Findings leading to discontinuation of the 2.5 mg/day estrogen group. J.Am.Med.Ass. 226: 652-7, 1973

85. Coronary Drug Project Research Group: The coronary drug project: Findings leading to discontinuation of the 2.5 mg/day estrogen group. J.Am.Med.Ass. 226: 652-7, 1973

86. Crawford BAL, Liu PY, Kean MT, Bleasel JF, Handelsman DJ: Randomized placebo-controlled trial of androgen effects on muscle and bone in men requiring long-term systemic glucocorticoid treatment. J.Clin. Endocrinol.Metab. 88: 3167-3176, 2003

87. Croxson TS, Chapman WE, Miller LK, Levit CD, Senie R, Zumoff B: Change in the hypothalamic-pituitary-gonadal axis in human immunodeficiency virus-infected homosexual men. J.Clin.Endocrinol.Metab. 68: 317-21, 1989

88. Cummings DE, Kumar N, Bardin CW, Sundaram K, Bremner WJ: Prostate-sparing effects in primates of the potent androgen 7α-methyl-19-nortestosterone: a po-

tential alternative to testosterone for androgen replacement and male contraception. J.Clin.Endocrinol.Metab. 83: 4212-9, 1998

89. Cunnah D, Besser M: Management of prolactinomas. Clin.Endocrinol. 34: 231-5, 1991

90. Cunningham GR, Cordero E, Thornby JI: Testosterone replacement with transdermal therapeutic systems. J.Am.Med.Ass. 261: 2525-30, 1989

91. Cunningham GR: Overview of androgens on the normal and abnormal prostate. In: Bhasin S, Gabelnick HL, Spieler JM, Swerdloff RS, Wang C, Kelly C: Pharmacology, biology, and clinical applications of androgens. John Wiley, New York, 79-93, 1996

92. Cunningham GR: Overview of androgens on the normal and abnormal prostate. In: Bhasin S, Gabelnick HL, Spieler JM, Swerdloff RS, Wang C, Kelly C: Pharmacology, Biology, and Clinical Applications of Androgens. John Wiley, New York, 79-93, 1996

93. Danner C, Frick J: Androgen substitution with testosterone containing nasal drops. Int.J.Androl. 3: 429-35, 1980

94. Davidson JM, Kwan M, Greenleaf WJ: Hormonal replacement and sexuality in men. Clinics in Endocrinology and Metabolism 11: 599-623, 1982

95. De Kretser DM, McLachlan RI, Robertson DM, Wreford NG: Control of spermatogenesis by follicle stimulating hormone and testosterone. Bailliere's Clin.Endocrinol.Metab. 6: 335-54, 1992

96. de la Chapelle A, Page DC, Brown L, Kaski U, Parvinen T, Tippett PA: The origin of 45,X males. Am.J.Hum.Genet. 38: 330-40, 1986

97. Deladoey J, Flück C, bex M, Yoshimura N, Harada N, Mullis PE: Aromatase deficiency caused by a novel $P450_{arom}$ gene mutation: impact of absent estrogen production on serum gonadotropin concentration in a boy. J.Clin.Endocrinol.Metab. 84: 4050-4, 1999

98. Dequeker J, Geusens P: Anabolic steroids and osteoporosis. Acta Endocrinol. 271 (Suppl.): 45-52, 1985

99. Deslypere JP, Young M, Wilson JD, McPhaul MJ: Testosterone and 5α-dihydrotestosterone interact differently with the androgen receptor to enhance transcription of the MMTV-CAT reporter gene. Mol.Cell.Endocrinol. 88: 15-22, 1992

100. Dobs AS, Dempsey MA, Ladenson PW, Polk BF: Endocrine disorders in men infected with human immunodeficiency virus. Am.J.Med. 84: 611-6, 1988

101. Drop SLS, de Waal WJ, De Muinck Keizer-Schrama SMPF: Sex steroid treatment of constitutionally tall stature. Endocr.Rev. 19: 540-58, 1998

102. Eagon PK, Elm MS, Stafford EA, Porter LE: Androgen receptor in human liver: characterization and quantitation in normal and diseased liver. Hepatology 19: 92-100, 1994

103. Eakman GD, Dallas JS, Ponder SW, Keenan BS: The effects of testosterone and dihydrotestosterone on hypothalamic regulation of growth hormone secretion. J.Clin.Endocrinol.Metab. 81: 1217-23, 1996

104. Eggert-Kruse W, Rohr G, Jochum R, Adolph M, Runnebaum B: Einfluß von Schwermetallen auf die In-vitro-Interaktion zwischen menschlichen Spermien und Zervixschleim. Dtsch.med.Wschr. 117: 1383-9, 1992

105. Evans DB, Crichlow RW: Carcinoma of the male breast and Klinefelter's syndrome: is there an association. Ca.Cancer J.Clin. 37: 246-51, 1987

106. Faiman C, Winter JSD, Reyes FI: Endocrinology of the fetal testis. In: Burger HG, De Kretser DM: The testis. Raven Press, New York, 81-105, 1981

107. Farrer JH, Walker AH, Rajfer J: Management of the postpubertal cryptorchid testis: a statistical review. J.Urol. 134: 1071-6, 1985

108. Fechner P, Marcantonio SM, Jaswaney V, et al: The role of the sex-determining region Y gene in the etiology of 46,XX maleness. J.Clin.Endocrinol.Metab. 76: 690-5, 1993

109. Feldman HA: Age trends in the levels of serum testosterone and other hormones in the middle-aged men: longitudinal results from the Massachusetts Male Aging Study. J.Clin.Endocrinol.Metab. 87: 589-98, 2002

110. Ferro P, Catalano MG, Dell ER, Fortunati N, Pfeffer U: The androgen receptor CAG repeat: a modifier of carcinogenesis? Mol.Cell.Endocrinol. 193: 109-20, 2002

111. Finkelstein JS, Klibanski A, Neer RM, et al: Increases in bone density during treatment of men with idiopathic hypogonadotropic hypogonadism. J.Clin.Endocrinol.Metab. 69: 776-83, 1989

112. Finkelstein JS, Neer RM, Biller BMK, Crawford JD, Klibanski A: Osteopenia in men with a history of delayed puberty. N.Engl.J.Med. 326: 600-4, 1992

113. Finkelstein JS, O'Dea LSTL, Whitcomb RW, Crowley WF: Sex steroid control of gonadotropin secretion in the human male. II. Effects of estradiol administration in normal and gonadotropin-releasing hormone-deficient men. J.Clin.Endocrinol.Metab. 73: 621-8, 1991

114. Finkelstein JS, Whitcomb RW, O'Dea LSTL, Longcope C, Schoenfeld DA, Crowley WF: Sex steroid control of gonadotropin secretion in the human male. I. Effects of testosterone administration in normal and gonadotropin-releasing hormone-deficient men. J.Clin.Endocrinol.Metab. 73: 609-20, 1991

115. Finkelstein JS: Androgens and osteoporosis: clinical aspects. In: Bhasin S, Gabelnick HL, Spieler JM, Swerdloff RS, Wang C, Kelly C: Pharmacology, biology, and clinical applications of androgens. John Wiley, New York, 265-77, 1996

116. Franchi F, Luisi M, Kicovic PM: Long-term study of oral testosterone undecanoate in hypogonadal males. Int.J.Androl. 1: 270-8, 1978

117. Fried W, Morley C: Effects of androgenic steroids on erythropoiesis. Steroids 46: 799-826, 1985

118. Friedler S, Raziel A, Strassburger D, Schachter M, Bern O, Ron-El R: Outcome of ICSI using fresh and cryopreserved-thawed testicular spermatozoa in patients with non-mosaic Klinefelter's syndrome. Hum.Reprod. 16: 2616-20, 2001

119. Geller J: Androgen inhibition and BPH. In: Bhasin S, Gabelnick HL, Spieler JM, Swerdloff RS, Wang C, Kelly C: Pharmacology, biology, and clinical applications of androgens. John Wiley, New York, 103-10, 1996

120. George FW, Wilson JD: Hormonal control of sexual development. Vitam.Horm. 43: 145-96, 1986

121. Gharib SD, Wierman ME, Shupnik MA, Chin WW: Molecular biology of pituitary gonadotropins. Endocr.Rev. 11: 177-99, 1990

122. Giovannucci E, Platz EA, Stampfer MJ, Chan A, Krithivas K, Kawachi I, Willett WC, Kantoff PW. The CAG repeat within the androgen receptor gene and benign prostatic hyperplasia. Urology 53:121-5, 1999

123. Giovannucci E, Stampfer MJ, Chan A, Krithivas K, Gann PH, Hennekens CH, Kantoff PW: CAG repeat within the androgen receptor gene and incidence of surgery for benign prostatic hyperplasia in U.S. physicians. Prostate 39:130-4, 1999

124. Giri S, Thompson PD, Taxel P, et al: Oral estrogen improves serum lipids, homocysteine and fibrinolysis in elderly men. Atherosclerosis 137: 359-66, 1998

125. Gooren LJG, Polderman KH: Safety aspects of androgen therapy. In: Nieschlag E, Behre HM: Testosterone. Springer Verlag, Berlin, 182-203, 1990

126. Gooren, L. J. G.A ten-year safety study of the oral androgen testosterone undecanoate. J.Androl. 15: 212-15, 1994

127. Gordon DL, Krmpotic E, Thomas W, Gandy HM, Paulsen CA: Pathologic testicular findings in Klinefelter's syndrome 47,XXY vs 46,XY/47,XXY. Arch.Int.Med. 130: 726-9, 1972

128. Gorski RA: Androgen: effects on the brain: cognitive, sexual, and aggressive behavior, an overview. In: Bhasin S, Gabelnick HL, Spieler JM, Swerdloff RS, Wang C, Kelly C: Pharmacology, Biology, and Clinical Applications of Androgens. John Wiley, New York, 157-8, 1996

129. Gorski RA: Androgens and sexual differentiation of the brain. In: Bhasin S, Gabelnick HL, Spieler JM, Swerdloff RS, Wang C, Kelly C: Pharmacology, Biology, and Clinical Applications of Androgens. John Wiley, New York, 159-68, 1996

130. Gray A, Berlin JA, McKinlay JB, Longcope C: An examination of research design effects on the association of testosterone and male aging: results of a meta-analysis. J.Clin.Epidemiol. 44: 671-84, 1991

131. Gray A, Feldman HA, McKinlay JB, Longcope C: Age, disease, and changing sex hormone levels in middle-aged men: results of the Massachusetts male aging study. J.Clin.Endocrinol.Metab. 73: 1016-25, 1991

132. Greenspan SL, Oppenheim DS, Klibanski A: Importance of gonadal steroids to bone mass in men with hyperprolactinemic hypogonadism. Ann.Intern.Med. 110: 526-31, 1989

133. Griffin JE: Androgen resistance - the clinical and molecular spectrum. N.Engl.J.Med. 326: 611-8, 1992

134. Griggs RC, Kingston W, Jozefowicz RF, Herr BE, Forbes G, Halliday D: Effect of testosterone on muscle mass and muscle protein synthesis. J.Appl.Physiol. 66: 498-503, 1989

135. Grino PB, Griffin JE, Cushard WG, Wilson JD: A mutation of the androgen receptor associated with partial androgen resistance, familial gynecomastia, and fertility. J.Clin.Endocrinol.Metab. 66: 754-7761, 1988

136. Grino PB, Griffin JE, Wilson JD: Testosterone at high concentrations interacts with the human androgen receptor similarly to dihydrotestosterone. Endocrinology 126: 1165-72, 1990

137. Grinspoon S, Corcoran C, Askari H, et al: Effects of androgen administration in men with the AIDS wasting syndrome. Ann.Intern.Med. 129: 18-26, 1998

138. Grohe C, Kahlert S, Lobbert K: Cardiac myocytes and fibroblasts contain functional estrogen receptors. FEBS Letters 416: 107-12, 1997

139. Grumbach MM: Genetic mechanisms of sexual development. In: Vallet HL, Porter ICH: Genetic mechanisms of sexual development. Academic Press, New York, 33-73, 1979

140. Grundy SM, Cleeman JI, Daniels SR, Donato KA, Eckel RH, Franklin BA, Gordon DJ, Krauss RM, Savage PJ, Smith SC, Spertus JA, Costa F. American Heart Association; National Heart, Lung, and Blood Institute. Diagnosis and management of the metabolic syndrome: an American Heart Association/National Heart, Lung, and Blood Institute Scientific Statement. Circulation 112: 2735-2752, 2005

141. Hadziselimovic F: Histology and ultrastructure of normal and cryptorchid testes. In: Hadziselimovic F: Cryptorchidism. Springer-Verlag, Berlin, Heidelberg, New York, 35-58, 1983

142. Haffner SM, Shaten J, Stern MP, Smith GA, Kuller LH: Low levels of sex hormone-binding globulin and testosterone predict the development of non-insulin-dependent diabetes mellitus in men. Am.J.Epidemiol. 143: 889-97, 1996

143. Hajjar RR, Kaiser FE, Morley JE: Outcome of long-term testosterone replacement in older hypogonadal males: a retrospective analysis. J.Clin.Endocrinol.Metab. 82: 3793-6, 1997

144. Hak AE, Wittemann JC, De Jong FH, Geerlings MI, Hofman A, Pols HAP: Low levels of endogenous androgens increase the risk of atherosclerosis in elderly men: the Rotterdam study. J.Clin.Endocrinol.Metab. 87: 3632-9, 2002

145. Handelsman DJ, Conway AJ, Boylan LM, Turtle JR: Young's syndrome: obstructive azoospermia and chronic sinopulmonary infections. N.Engl.J.Med. 310: 3-9, 1984

146. Harari O, Bourne H, Baker G, Gronow M, Johnston I: High fertilization rate with intracytoplasmic sperm injection in mosaic Klinefelter's syndrome. Fertil.Steril. 63: 182-4, 1995

147. Hardelin JP, Levilliers J, Young J, et al: Xp22.3 deletions in isolated familial Kallmann's syndrome. J.Clin.Endocrinol.Metab. 76: 827-31, 1993

148. Harman SM, Metter EJ, Tobin JD, Pearson JD, Blackman MR: Longitudinal effects of aging on serum total and free testosterone levesl in healthy males. J.Clin.Endocrinol.Metab. 86: 724-31, 2001

149. Heinemann LAJ, Zimmermann T, Vermeulen A, Thiel C, Hummel W: A new aging males' symptoms rating scale. Ageing Male 2: 105-14, 2001

150. Heinlein CA, Chang C: Androgen receptor (AR) coregulators: an overview. Endocr.Rev. 23: 175-200, 2002

151. Henriksson P, Edhag O, Eriksson A, Johansson SE: Patients at high risk of cardiovascular complications in oestrogen treatment of prostatic cancer. Br.J.Urol. 63: 186-90, 1989

152. Herrmann BL, Saller B, Janssen OE, Gocke P, Bockisch A, Sperling H, Mann K, Broecker K: Impact of estrogen replacement therapy in a male with congenital aromatase deficiency caused by a novel mutation in the CYP19 gene. J.Clin.Endocrinol.Metab. 87: 5476-84, 2002

153. Herzog AG, Klein P, Jacobs AR: Testosterone versus testosterone and testolactone in treating reproductive and sexual dysfunction in men with epilepsy and hypogonadism. Neurology 50: 782-4, 1998

154. Hiraga T, Shimokawa K, Murase T, Yokoyama M: Reduction of serum lipoprotein(a) by estrogen in men with prostatic cancer. Endocr.J. 40: 507-13, 1993

155. Hodges LK, Tung L, Graham JD, Yan XD, Horwitz KB, Horwitz LD: Heterogeneity of estrogen receptor expression and function in human vascular smooth muscle. Circulation 98: I-799, 1998

156. Holmäng S, Marin P, Lindstedt G, Hedelin H: Effect of long-term oral testosterone undecanoate treatment on prostate volume and serum prostate-specific antigen concentration in eugonadal middle-aged men. Prostate 23: 99-106, 1996

157. Hryb DJ, Khan MS, Romas NA, Rosner W: The control of the interaction of sex hormone-binding globulin with its receptor by steroid hormones. J.Biol.Chem. 265: 6048-54, 1990

158. Hubert W: Psychotropic effects of testosterone. In: Nieschlag E, Behre HM, eds. Testosterone. Springer Verlag, Berlin, 51-71, 1990

159. Hulley S, Grady D, Bush T: Randomized trial of estrogen plus progestin for secondary prevention of coronary heart disease in postmenopausal women. Heart and estrogen/progestin replacement study (HERS) Research Group. J.Am.Med.Ass. 280: 605-13, 1998

160. Hutson JM, Donahoe PK: The hormonal control of testicular descent. Endocr.Rev. 7: 270-83, 1986

161. International Diabetes Federation. Worldwide definition of the metabolic syndrome. (www.idf.org/webdata/docs/IDF_Metasyndrome_definition.pdf) 2005

162. International Prostate Health Council Study Group: Estrogens and prostatic disease. Prostate 45: 87-100, 2000

163. Isaacs JT: Role of androgens in normal and malignant growth of the prostate. In: Bhasin S, Gabelnick HL, Spieler JM, Swerdloff RS, Wang C, Kelly C: Pharmacology, biology, and clinical applications of androgens. John Wiley, New York, 95-101, 1996

164. Isaacs JT: Role of androgens in prostatic cancer. Vitam.Horm. 49: 433-502, 1994

165. Ishak KG: Hepatic lesions caused by anabolic and contraceptive steroids. Sem.Liver Dis. 1: 116-28, 1981

166. Isidori AM, Giannetta E, Greco EA, Gianfrilli D, Bonifacio V, Isidori A, Lenzi A, Fabbri A. Effects of testosterone on body composition, bone metabolism and serum lipid profile in middle-aged men: a meta-analysis. Clin Endocrinol 63:280-93, 2005

167. Itil TM, Michael ST, Shapiro DM, Itil KZ: The effects of mesterolone, a male sex hormone in depressed patients (a double blind controlled study). Meth. Find. Exp.Clin.Pharmacol. 6: 331-7, 1984

168. Jackson JA, Kleerekoper M, Parfitt AM, Rao DS, Villanueva AR, Frame B: Bone histomorphometry in hypogonadal and eugonadal men with spinal osteoporosis. J.Clin.Endocrinol.Metab. 65: 53-8, 1987

169. Jackson JA, Kleerekoper M: Osteoporosis in men: diagnosis, pathophysiology, and prevention. Medicine 69: 137-52, 1990

170. Jackson JA, Riggs MW, Spiekerman AM: Testosterone deficiency as a risk factor for hip fractures in men: a case-control study. Am.J.Med.Sci. 304: 4-8, 1992

171. Jackson JA, Waxman J, Spiekerman M: Prostatic complications of testosterone replacement therapy. Arch.Int.Med. 149: 2365-6, 1989

172. Jaffe MD: Effect of testosterone cypionate on postexercise ST segment depression. Br.Heart J. 39: 1217-22, 1977

173. Jain P, Rademaker AW, McVary KT: Testosterone supplementation for erectile dysfunction: results of a meta-analysis. J.Urol. 164: 371-5, 2001

174. Jakob, F.; Siggelkow, H.; Homann, D.; Köhrle, J.; Adamski, J.; Schütze, N. Local estradiol metabolism in osteoblast- and osteoclast-like cells. Journal of Steroid

Biochemistry and Molecular Biology 61, 167-174. 1997. Ref Type: Journal (Full)

175. Janowsky DS, Oviatt SK, Orwoll ES: Testosterone influences spatial cognition in older men. Behav.Neurosci. 108: 325-32, 1994

176. Jegou B: The Sertoli cell. Bailliere's Clin.Endocrinol.Metab. 6: 273-311, 1992

177. Jockenhövel F, Bals-Pratsch M, Bertram HP, Nieschlag E: Seminal lead and copper in fertile and infertile men. andrologia 22: 503-11, 1990

178. Jockenhövel F, Behre HJM, Nieschlag E: Therapie des Hypogonadismus und der Infertilität des Mannes. Med.Klin. 85: 87-91 (Teil I), 145-50 (Teil II) 212-5 (Teil III), 1990

179. Jockenhövel F, Blum WF, Vogel E, et al: Testosterone substitution normalizes elevated leptin serum levels in hypogonadal men. J.Clin.Endocrinol.Metab. 82: 2510-3, 1997

180. Jockenhövel F, Bullmann C, Schubert M, et al: Influence of various modes of androgen substitution on serum lipids and lipoproteins in hypogonadal men. Metabolism 48: 590-6, 1999

181. Jockenhövel F, Gerhardts S: Erkrankungen von Hypothalamus und Hypophyse. Uni-Med Verlag, Bremen, 2002

182. Jockenhövel F, Gräwe A, Nieschlag E: A portable digital data recorder for long-term monitoring of scrotal temperatures. Fertil.Steril. 54: 694-700, 1990

183. Jockenhövel F, Krüsemann C, Jaeger A, Olbricht Th, Reinwein D: Vergleichbarkeit der 10 gebräuchlichsten kommerziell erhältlichen Radioimmunoassays und eines neuen Enzym-Immunoassays zur Bestimmung von Testosteron im Serum. Klin.Lab. 38: 81-8, 1992

184. Jockenhövel F, Lerchl A, Allolio B: Hormone gegen das Altern - Möglichkeiten und Grenzen. Dt.Ärztebl. 98: A2041-A2045, 2001

185. Jockenhövel F, Nieschlag E: Männliche Fertilitätsstörungen und Hypogonadismus: Hypothalamus. In: Hesch RD: Endokrinologie. Urban & Schwarzenberg, München, 832-39, 1989

186. Jockenhövel F, Nieschlag E: Primäre testikuläre Erkrankungen. In: Hesch RD: Endokrinologie. Urban & Schwarzenberg, München, 928-51, 1989

187. Jockenhövel F, Reinwein D: Klinefelter-Syndrom: Neue Erkenntnisse zur Klinik und Therapie. Dtsch.med.Wschr. 117: 383-9, 1992

188. Jockenhövel F, Swerdloff RS: Alterations in the steroidogenic capacity of Leydig cells in cryptorchid testis. In: Abney TO, Keel BA: The Cryptorchid Testis. CRC Press, Boca Raton, 35-54, 1989

189. Jockenhövel F, Vogel E, Kreutzer M, Reinhardt W, Lederbogen S, Reinwein D: Pharmacokinetics and pharmacodynamics of subcutaneous testosterone implants in hypogonadal men. Clin.Endocrinol. 45: 61-71, 1996

190. Jockenhövel F, Vogel E, Reinhardt W, Reinwein D: Effects of various modes of androgen substitution therapy on erythropoiesis. Eur.J.Med.Res. 2: 293-8, 1997

191. Jockenhövel F: Androgenmangel des älteren Mannes - Was bringt die Testosteron-Substitution? Dtsch.med.Wschr. 126: 247-52, 2001

192. Jockenhövel F: Androgen-Resistenz-Syndrom. In: Allolio B, Herrmann J, Olbricht Th, Rudorff KH, Schulte HM: 2. Intensivkurs für klinische Endokrinologie. pmi-Verlag, Frankfurt, 154-170, 1993

193. Jockenhövel F: Erektile Dysfunktion. In: Reinwein D: Endokrinologische Entscheidungen: klinische Algorithmen. Kohlhammer Verlag, Stuttgart, 15-7, 1994

194. Jockenhövel F: Hypogonadismus und Infertilität als Folge von allgemeinen Erkrankungen und Toxinen. Internist 34: 741-55, 1993

195. Jockenhövel F: Männlicher Hypogonadismus. In: Allolio B, Schulte HM: Praktische Endokrinologie. Urban & Schwarzenberg, München, 361-80, 1996

196. Jockenhövel F: Was ist gesichert in der Diagnostik und Therapie des partiellen Androgendefizits (PADAM). Urologe [B] 41: 325-30, 2001

197. Johnsen SG, Bennett EP, Jensen VG: Therapeutic effectiveness of oral testosterone. Lancet 2: 1473-5, 1974

198. Johnson VP, Aceto T, McMillin JM: Gonadotropin-induced fertility in a man with congenital hypogonadism after prolonged prior testosterone therapy. S.Dak.J.Med. 41: 5-7, 1988

199. Jordan WP: Allergy and topical irritation associated with transdermal testosterone administration: a comparison of scrotal and nonscrotal transdermal systems. Am.J.Contact Derm. 8: 108-13, 1997

200. Jorgensen JOL, Vahl N, Hansen TB, Fisker S, Hagen C, Christiansen JS: Influence of growth hormone and androgens on body composition in adults. Horm.Res. 45: 94-8, 1996

201. Josso N: Anti-müllerian hormone. Clin.Endocrinol. 25: 331-45, 1986

202. Kahn SM, Hryb DJ, Nakhla AM, Romas NA, Rosner W: Beyond carrier proteins: sex hormone binding globulin is synthesized in target cells. J.Endocrinol. 175: 113-20, 2000

203. Kaiser E, Kies N, Maas G, Schmidt H, Beach RC, Bormacher K, Herrmann W, and Richter E. The measurement of the psychotropic effects of an androgen in aging males with psychovegetative symptomatology: a controlled double blind study mesterolone vs. placebo. Neuro-Psychopharmacology 2, 505-15. 1978.

204. Kalinchenko SY, Kozlov GI, Gontcharov NP, Katsiya GH: Oral testosterone undecanoate reverses erectile dysfunction associated with diabetes mellitus in patients failing on sildenafil citrate therapy alone. Aging Male 6: 94-99, 2003

205. Kandeel FR, Swerdloff RS: Role of temperature in regulation of spermatogenesis and the use of heating as a method for contraception. Fertil.Steril. 49: 1-23, 1988

206. Kapoor D, Goddwin E, Channer KS, Jones TH: Testosterone replacement therapy improves insulin resistance, glycaemic control, visceral adiposity and hypercholesterolaemia in hypogonadal men with type 2 diabetes. Eur J Endocrinol 154: 899-906, 2006

207. Katznelson L, Finkelstein JS, Schoenfeld DA, Rosenthal DI, Anderson EJ, Klibanski A: Increase in bone density and lean body mass during testosterone administration in men with acquired hypogonadism. J.Clin.Endocrinol.Metab. 81: 4358-65, 1996

208. Kenny AM, McGee D, Joseph C, Covault J, Abreu C, Raisz LG. Lack of association between androgen receptor polymorphisms and bone mineral density or physical function in older men. Endocr Res 31:285-93, 2005

209. Kerr JB: Functional cytology of the human testis. Bailliere's Clin.Endocrinol.Metab. 6: 235-50, 1992

210. Khaw KT, Barrett-Connor E: Lower endogenous androgens predict central adiposity in men. Ann.Epidemiol. 2: 675-82, 1992

211. Khosla S.:Oestrogen, bones and men: when testosterone just isn't enough. Clin Endocrinol 56:291-3, 2002

212. Kikuchi TA, Skowsky WR, El-Toraei I, Swerdloff RS: The pituitary-gonadal axis in spinal cord injury. Fertil.Steril. 27: 1142-5, 1976

213. King RJB: Effects of steroid hormones and related compound on gene transcription. Clin.Endocrinol. 36: 1-14, 1992

214. Kloer H, Hoogen H, Nieschlag E: Trial of high-dose testosterone undecanoate in treatment of male infertility. Int.J.Androl. 3: 121-9, 1980

215. Klotz L, Correia A, Zhang W: The relationship between the androgen receptor CAG repeat polymorphism length and the response to intermittent androgen suppression therapy for advanced prostate cancer. Prostate Cancer Prostatic Dis 8:179-83, 2005

216. Klotz T, Hurrelmann K, Eickenberg HU: Der frühe Tod des starken Geschlechts. Dt.Ärztebl. 95: A460-64 1998

217. Knudtzon J, Aarskog D: 45,X/46,XY mosaicism. Eur.J.Pediatr. 146: 266-71, 1987

218. Korenman SG, Morley JE, Mooradian AD, Stanik Davis S, Kaiser FE, Silver AJ, Viosca SP, Garza D: Secondary hypogonadism in older men: its relation to impotence. J.Clin.Endocrinol.Metab. 71: 963-9, 1990

219. Koyama Y, Iigaya T, Saito S: Tuberculous epididymo-orchitis. Urol. 4: 419-21, 1988

220. Krane RJ, Goldstein I, de Tajeda IS: Impotence. N.Engl.J.Med. 321: 1648-59, 1989

221. Krieg M, Tunn S: Androgens and human benign prostatic hyperplasia (BPH). In: Nieschlag E, Behre HM: Testosterone. Springer Verlag, Berlin, 219-44, 1990

222. Krieger JN, Coombs RW, Collier AC, et al: Fertility parameters in men infected with human immunodeficiency virus. J.Infect.Dis. 164: 464-9, 1991

223. Krishnaswamy V, Kumarasamy T, Venkatesan V, Shroff S, Jayanth VR, Paul SF. South Indian men with reduced CAG repeat length in the androgen receptor gene have an increased risk of prostate cancer.J Hum Genet 51:254-7, 2006.

224. La Spada AR, Wilson EM, Lubahn DB, Harding AE, Fischbeck KH: Androgen receptor gene mutations in X-linked spinal and bulbar muscular atrophy. Nature 352: 77-9, 1991

225. Laaksonen DE, Niskanen L, Punnonen K, Nyysonen K, Tuomainen T-P, Valkonen VP, Salonen R, Salonen JT: Testosterone and sex hormone-binding globulin predict the metabolic syndrome and diabetes in middle-aged men. Diabetes Care 27:1036-1041, 2004

226. Lamberts SWJ, van den Beld AW, van der Lely AJ: The endocrinology of aging. Science 278: 419-24, 1998

227. Largo RH, Prader A: Pubertal development in Swiss boys. Helv.Paediat.Acta 38: 211-28, 1983

228. Larsson B, Svärdsudd K, Welin L, Wilhelmsen L, Björntorp P, Tibblin G: Abdominal adipose tissue distribution, obesity, and risk of cardiovascular disease and death: 13 year follow up of participants in the study of men born in 1913. Br.Med.J. 288: 1401-4, 1984

229. Lauber ME, Sarasin A, Lichtensteiger W: Sex differences and androgen-dependent regulation of aromatase (CYP19) mRNA expression in the developing and adult rat brain. J.Steroid Biochem.Mol.Biol. 61: 359-64, 1997

230. Layman LC, Cohen DP, Jin M, Xie J, Li Z, Reindollar R, Bolbolan S, Bick DP, Sherins R R, Duck LW, Musgrove LC, Sellers JC, Neill JD: Mutations in gonadotropin-releasing hormone receptor gene cause hypogonadotropic hypogonadism. Nature Genetics 18: 14-15, 1998.

231. Lee MM, Donahoe PK: Mullerian inhibiting substance: a gonadal hormone with multiple functions. Endocr.Rev. 14: 152-64, 1993

232. Leifke E, Gorenoi V, Wichers C, von zur Mühlen A, von Büren E, Brabant G: Age-related changes of serum sex hormones, insulin like growth factor-1 and sex-hormone binding globulin levels in men: cross sectional data from a healthy male cohort. Clin.Endocrinol. 53: 689-95, 2000

233. Levine AC: Pathogenesis and medical treatment management of benign prostatic hyperplasia. Trends Endocrinol.Metab. 6: 128-32, 1995

234. Levis WR, Lanzan AP, Swersie S, Meeker HC, Schuller-Levis GB, Bardin CW: Testicular dysfunction in leprosy: relationships of FSH, LH and testosterone to disease classification, activity and duration. Lepr.Rev. 60: 94-101, 1989

235. Liang T: Androgens and skin: 5α-reductase activity in skin and androgen dependent skin disorders. In:

Bhasin S, Gabelnick HL, Spieler JM, Swerdloff RS, Wang C, Kelly C, eds. Pharmacology, biology, and clinical applications of androgens. John Wiley, New York, 247-56, 1996

236. Lindner V, Kim SK, Karas RH, Kuiper GGJM, Gustaffsson JA, Mendelsohn ME: Increased expression of estrogen receptor-beta mRNA in male blood vessels after vascular injury. Circ.Res. 83: 224-9, 1998

237. Lindstedt G, Nyström E, Matthews C, Ernest I, Janson PO, Chatterjee VKK: Follitropin (FSH) deficiency in an infertile male due to FSH beta gene mutation. Clin.Chem.Lab.Med. 36: 663-5, 1998

238. MacLean HE, Warne GL, Zajac JD: Intersex disorders: shedding light on male sexual differentiation beyond SRY. Clin.Endocrinol. 46: 101-8, 1997

239. Maes M, Lee P A, Jeffs RD, Sultan C, Migeon CJ: Phenotypic variation in a family with partial androgen insensitivity syndrome. Am.J.Dis.Child. 134: 470-3, 1980

240. Marcelli M, Tilley WD, Zoppi S, Griffin JE, Wilson JD, McPhaul MJ: Molecular basis of androgen resistance. J.Endocrinol.Invest. 15: 149-59, 1992

241. Marin P, Andersson B, Ottosson M, Olbe L, Chodhury B, Kvist H, Holm G, Sjöström L, Björntorp P: The morphology and metabolism of intraabdominal adipose tissue in men. Metabolism 41: 1242-8, 1992

242. Marin P, Arver S: Androgens and abdominal obesity. Bailliere's Clin.Endocrinol.Metab. 12: 441-51, 1998

243. Marin P, Holmäng S, Gustafsson C, et al: Androgen treatment of abdominally obese men. Obesity Res. 1: 245-51, 1993

244. Marin P, Holmäng S, Jönsson L, Sjöström L, Kvist H, Holm G, Lindstedt G, Björntorp P: The effects of testosterone treatment on body composition and metabolism in middle-aged obese men. Int.J.Obesity 16: 991-7, 1992

245. Marin P, Lönn L, Andersson B, Oden B, Olbe L, Bengtsson BA, Björntorp P: Assimilation of triglycerides in subcutaneous and intraabdominal adipose tissue in vivo in men: effects of testosterone. J.Clin.Endocrinol.Metab. 81: 1018-22, 1996

246. Marin P, Oden B, Björntorp P: Assimilation and metabolism of triglycerides in subcutaneous abdominal and femoral adipose tissue in vivo in men: effects of androgens. J.Clin.Endocrinol.Metab. 80: 239-43, 1995

247. Marin P: Testosterone and regional fat distribution. Obes.Res. 3: 609S-12S, 1995

248. Marmorston J, Moore FJ, Kuzma OT, Magidson O, Weiner J: Effect of premarin on survival in men with myocardial infarction. Proc.Soc.Exp.Biol.Med. 105: 618-20, 1960

249. Matsumoto AM, Sandblom RE, Schoene RB, et al: Testosterone replacement in hypogonadal men: effects on obstructive sleep apnoea respiratory drives, and sleep. Clin.Endocrinol. 22: 713-21, 1985

250. Mauras N: Growth hormone and sex steroids: interactions in puberty. Endocrinol.Metab.Clin.North Am. 30: 529-44, 2001

251. Mbanya JCN, Mendelow AD, Crawford PJ, Hall K, Dewar JH, Kendall-Taylor P: Rapid resolution of visual abnormalities with medical therapy alone in patients with large prolactinomas. Br.J.Neurosurg. 7: 519-27, 1993

252. McCulloch DK, Young RJ, Prescott RJ, Campbell IW, Clarke BF: The natural history of impotence in diabetic men. Diabetol. 26: 437-40, 1984

253. McElreavey K, Fellous M: Sex determination and the Y chromosome. Am.J.Med.Gen. 89: 176-85, 1999

254. McLaughlin T, Abbasi F, Cheal K, Chu J, Lamendola C, Reaven G. Use of metabolic markers to identify overweight individuals who are insulin resistant. Ann Intern Med 139:802-809, 2003

255. McMahon DR, Kramer SA, Husmann DA: Micropenis: does early treatment with testosterone do more harm than good? J.Urol. 154: 825-9, 1995

256. Meikle AW, Arver S, Dobs AS, Sanders SW, Lakshminaryan R, Mazer NA: Pharmacokinetics and metabolism of a permeation-enhanced testosterone transdermal system in hypogonadal men: influence of application site - a clinical research center study. J.Clin.Endocrinol.Metab. 81: 1832-40, 1996

257. Meikle AW, Arver S, Dobs AS, Sanders SW, Mazer NA: Androderm: a permeation enhanced non-scrotal testosterone transdermal system for the treatment of male hypogonadism. In: Bhasin S, Gabelnick HL, Spieler JM, Swerdloff RS, Wang C, Kelly C: Pharmacology, biology, and clinical applications of androgens. John Wiley, New York, 449-57, 1996

258. Meikle AW, Mazer NA, Moellmer JF, et al: Enhanced transdermal delivery of testosterone across nonscrotal skin produces physiological concentrations of testosterone and its metabolites in hypogonadal men. J.Clin.Endocrinol.Metab. 74: 623-8, 1992

259. Migeon CJ, Brown TR, Lanes R, Palacios A, Amrhein JA, Schoen EJ: A clinical syndrome of mild androgen insensitivity. J.Clin.Endocrinol.Metab. 59: 672-8, 1984

260. Mikkola AK, Ruutu ML, Aro JL, Ranniko SA, Salo JO: Parenteral polyoestradiol phosphate vs orchidectomy in the treatment of advanced prostatic cancer. Efficacy and cardiovascular complications: a 2-year follow-up report of a national, prospective prostatic cancer study. Finnprostate Group. Br.J.Urol. 82: 63-8, 1998

261. Mitchell R, Hollis S, Rothwell C, Robertson WR: Age related changes in the pituitary-testicular axis in normal men; low testosterone results from decreased bioactive LH drive. Clin.Endocrinol. 42: 501-7, 1995

262. Moffatt A, Zonderman A, Metter EJ, Blackman MR, Harman SM, Resnick S: Longitudinal assessment of serum free testosterone concentration predicts memory

performance and cognitive status in elderly men. J.Clin.Endocrinol.Metab. 87: 5001-7, 2002

263. Mooradian AD, Morley JE, Korenman SG: Biological actions of androgens. Endocr.Rev. 8: 1-27, 1987

264. Morales A, Bain J, Ruijs A, Chapdelaine A, Tremblay RR: Clinical practice guidelines for screening and monitoring male patients receiving testosterone supplementation therapy. Int.J.Impot.Res. 8: 95-7, 1996

265. Morales A, Lunenfeld B: Investigation, treatment and monitoring of late-onset hypogonadism in males. Ageing Male 5: 74-86, 2002

266. Morgentaler A, Bruning CO, DeWolf WC: Occult prostate cancer in men with low serum testosterone levels. J.Am.Med.Ass. 276: 1904-6, 1996

267. Morgentaler A: Male impotence. Lancet 354: 1713-8, 1999

268. Morishima A, Grumbach MM, Simpson ER, Fisher C, Qin K: Aromatase deficiency in male and female siblings caused by a novel mutation and the physiological role of estrogens. J.Clin.Endocrinol.Metab. 80: 3689-98, 1995

269. Morita R, Yamamoto I, Fukunaga M, et al: Changes in sex hormones and calcium regulating hormones with reference to bone mass associated with aging. Endocrinol.Jpn. 26: 15-22, 1979

270. Morley JE, Kaiser FE: Testicular function in the aging male. In: Armbrecht HJ, Coe RM, Wongsurawat N: Endocrine function and aging. Springer Verlag, Berlin, 99-114, 1990

271. Morley JE, Perry HM, Kaiser FE, et al: Effects of testosterone replacement therapy in old hypogonadal males: a preliminary study. J.Am.Geriatr.Soc. 41: 149-52, 1993

272. Morley JE, Perry HM: Androgen deficiency in aging men: role of testosterone replacement therapy. J Lab Clin Med 135: 370-8, 2000

273. Morley JE: Impotence. Am.J.Med. 80: 897-905, 1986

274. Morrison JA, deGroot I, Edwards BK, et al: Plasma cholesterol and triglyceride levels in 6775 school children, ages 6 - 17. Metabolism 26: 1199-211, 1977

275. Morrison JA, Laskarzewski PM, Rauh JL, et al: Lipids, lipoproteins, and sexual maturation during adolescence: the Princeton Maturation Study. Metabolism 28: 641-9, 1979

276. Morrow AF, Gyorki S, Warne GL, et al: Variable androgen receptor levels in infertile men. J.Clin.Endocrinol.Metab. 64: 1115-21, 1987

277. Murray FT, Wyss HU, Thomas RG, Spevak M, Glaros AG: Gonadal dysfunction in diabetic men with organic impotence. J.Clin.Endocrinol.Metab. 65: 127-35, 1987

278. Nanjee MN, Wheeler MJ: Plasma free testosterone - is an index sufficient? Ann.Clin.Biochem. 22: 387-90, 1985

279. Netley C: Predicting intellectual functioning in 47,XXY boys from characteristics of sibs. Clin.Genet. 32: 24-7, 1987

280. Nielsen J, Sorensen K: The importance of early diagnosis of Klinefelter's syndrome. In: Bandmann HJ, Breit R: Klinefelter's Syndrome. Springer Verlag, Berlin, 170-87, 1984

281. Nieschlag E, Büchter D, von Eckardstein S, Abshagen K, Simoni M, Behre HM: Repeated intramuscular injections of testosterone undecanoate for substitution therapy in hypogonadal men. Clin.Endocrinol. 51: 757-63, 1999

282. Nieschlag E, Cüppers HJ, Wiegelmann W, Wickings EJ: Bioavailability and LH-suppressing effect of different testosterone preparations in normal and hypogonadal men. Horm.Res. 7: 138-45, 1976

283. Nieschlag E, Jockenhövel F: Behandlung der männlichen Infertilität mit Gonadotropinen. In: Lehmann F, Breckwoldt M: Gonadotropine. Enke Verlag, Stuttgart, 93-100, 1991

284. Nieschlag E, Jockenhövel F: Hypogonadismus beim Mann - Androgenmangel-Syndrom. In: Hesch RD: Endokrinologie. Urban & Schwarzenberg, München, 1216-20, 1989

285. Nieschlag E, von zur Mühlen A, Sippell W: Männliche Gonaden. In: Deutsche Gesellschaft für Endokrinologie: Rationelle Diagnostik in der Endokrinologie. Thieme Verlag, Stuttgart, 186-212, 1993

286. O'Carroll R, Bancroft J: Testosterone therapy for low sexual interest and erectile dysfunction in men: a controlled study. Br.J.Psychiatry 145: 146-51, 1984

287. O'Connell MJ, Ramsey HE, Whang-Peng J, Wiernik PH: Testicular feminization syndrome in three sibs: emphasis on gonadal neoplasia. Am.J.Med.Sci. 265: 321-33, 1973

288. O'Donnell L, Robertson KM, Jones ME, Simpson ER: Estrogen and spermatogenesis. Endocr.Rev. 22: 289-318, 2001

289. Oh JY, Barrett-Connor E, Wedick NM, Wingard DL; Rancho Bernardo Study. Endogenous sex hormones and the development of type 2 diabetes in older men and women: the Rancho Bernardo study.Diabetes Care. 25:55-60, 2002

290. Oliver MF, Boyd GS: Influence of reduction of serum lipids on prognosis of coronary heart disease. Lancet 2: 499-505, 1961

291. Ongphiphadhanakul B, Rajatanavin R, Chailurkit L, et al: Serum testosterone and its relation to bone mineral density and body composition in normal males. Clin.Endocrinol. 43: 727-33, 1995

292. Ongphiphadhanakul B, Thamprajamchit S, Chanprasertyothin S, Chailurkit L, Rajatanavin R: Effect of estrogen replacement on insulin sensitivity, serum lipid and bone resorption marker in hypogonadal males. Maturitas 42: 85-9, 2002

293. Orwoll ES, Klein RF: Osteoporosis in men. Endocr.Rev. 16: 87-116, 1995

294. Orwoll ES: Androgens as anabolic agents for bone. Trends Endocrinol.Metab. 7: 77-84, 1996

295. Page St, Amory JK, Bowman FD, Anawalt BD, Matsumoto AM, Bremner WJ, Tenover JL: Exogenous testosterone alone or with finasteride increases physical performance, grip strength, and lean body mass in older men with low serum testosterone. J. Clin. Endocrinol. Metab. 90: 1502-10, 2005

296. Pechersky AV, Mazurov VI, Semiglazov VF, Karpischenko Al, Mikhailichenko VV, Udentsev AV: Androgen administration in middle aged and ageing men: effects of oral testosterone undecanoate on dihydrotestosterone, oestradiol and prostate volume. Int.J.Androl. 25: 119-125, 2002

297. Perez-Palacios G, Ortiz S, Lopez-Amor E, et al: Familial incomplete virilization due to partial end organ insensitivity to androgens. J.Clin.Endocrinol.Metab. 41: 946-52, 1975

298. Perrin G, Treluyer C, Trouillas J, Sassolas G, Goutelle A: Surgical outcome and pathological effects of bromocriptine preoperative treatment in prolactinomas. Path.Res.Pract. 187: 587-92, 1991

299. Perry PJ, Yates WR, Williams RD, Anderson AE, Mclndoe,JH, Lund BC, Holman TL: Testosterone therapy in late-life major depression in males. J.Clin.Psychiatry 63: 1096-1101, 2002

300. Peterson RE, Imperato-McGinley J: Male pseudohermaphroditism due to inherited deficiencies of testosterone biosynthesis. In: Serio M, Motta M, Zanisi M, Martini L: Sexual differentiation. Basic and clinical aspects. Raven Press, New York, 301-19, 1984

301. Pfeilschifter J, D'Souza SM, Mundy GR: Effects of transforming growth factor-β on osteoblastic osteosarcoma cells. Endocrinology 121: 212-8, 1987

302. Phillip M, Arbelle JE, Segev Y, Parvari R: Male hypogonadism due to a mutation in the gene for the beta-subunit of follicle-stimulating hormone. N.Engl.J.Med. 338: 1729-32, 1998

303. Phillips GB, Castelli WP, Abbott RD, McNamara PM: Association of hyperestrogenemia and coronary heart disease in men in the Framingham cohort. Am.J.Med. 74: 863-9, 1983

304. Phillips GB, Pinkernell BH, Jing TY: The association of hyperestrogenemia with coronary thrombosis in men. Arterioscler.Thromb. 16: 1383-7, 1996

305. Phillips GB: Relationship of serum sex hormones to coronary heart disease. Steroids 58: 286-90, 1993

306. Pirke KM, Doerr P: Age related changes in free plasma testosterone, dihydrotestosterone and oestradiol. Acta Endocrinol. 80: 171-8, 1970

307. Pitteloud N, Mootha VK, Dwyer AA, Hardin M, Lee H, Eriksson KF, Tripathy D, Yialamas, M, Groop L, Elahi D, Hayes FJ: Relationship Between Testosterone Levels, Insulin Sensitivity, and Mitochondrial Function in Men. Diabetes Care 28:1636-1642, 2005

308. Pope HG, Cohane GH, Kanayama G, Siegel AJ, Hudson JI: Testosterone gel supplementation for men with refractory depression: a randomized, placebo-controlled trial. Am.J.Psychiatry 160: 105-111, 2003

309. Potashnik G, Yanai-Inbar I: Dibromochloropropane (DBCP): an 8-year reevaluation of testicular function and reproductive performance. Fertil.Steril. 47: 317-23, 1987

310. Pryor JL: Androgen supplementation causes BPH. In: Bhasin S, Gabelnick HL, Spieler JM, Swerdloff RS, Wang C, Kelly C: Pharmacology, biology, and clinical applications of androgens. John Wiley, New York, 111-14, 1996

311. Quinton R, Duke VM, Robertson A, Kirk JMW, Matfin G, de Zoysa PA, Azcona C, MacColl GS, Jacobs HS, Conway GS, Besser GM, Stanhope R, Bouloux PMG: Idiopathic gonadotrophin deficiency: genetic questions addressed through phenotypic characterization. Clin.Endocrinol. 55: 163-74, 2001

312. Rajfer J, Handelsman DJ, Swerdloff RS, et al: Hormonal therapy of cryptorchidism. N.Engl.J.Med. 314: 466-70, 1986

313. Reid IR, Wattie DJ, Evans MC, Stapleton JP: Testosterone therapy in glucocorticoid-treated men. Arch.Intern.Med. 156: 1173-7, 1996

314. Reis SE, Holubkov R, Zell KA: Conjugated estrogens acutely abolish abnormal cold-induced coronary vasoconstriction in male cardiac allografts. Circulation 97: 23-5, 1998

315. Reis SE, Holubkov R, Zell KA: Estrogen acutely abolishes abnormal cold-induced coronary constriction in men. Chest 114: 1556-61, 1998

316. Revelli A, Massabrio M, Tesabrik: Nongenomic actions of steroid hormones in reproductive tissues. Endocr.Rev. 19: 3-17, 1998

317. Riggs BL, Koshla S, Melton LJ: Sex Steroids and the Construction and Conservation of the Adult Skeleton. Endocr.Rev. 23: 279-302, 2002

318. Ringe JD, Dorst AJ: Osteoporose bei Männern. Dtsch.med.Wschr. 119: 943-7, 1994

319. Robertson DM, Risbridger GP, De Kretser DM: The physiology of testicular inhibin and related proteins. Bailliere's Clin.Endocrinol.Metab. 6: 355-72, 1992

320. Rolf C, von Eckardstein S, Koken U, Nieschlag E: Testosterone substitution of hypogonadal men prevents the age-dependent increases in body mass index, body fat

and leptin seen in healthy ageing men: results of a cross-sectional study. Eur.J.Endocrinol. 146: 505-11, 2002

321. Rosano GMC, Leonardo F, Pagnotta P, Pelliccia F, Panina G, Cerquetani E, della Monica PL, Bonfilgi B, Volpe M, Chierchia SL: Acute anti-ischemic effect of testosterone in men with coronary artery disease. Circulation 99: 1666-70, 1999

322. Rosenfeld RG, Northcraft GB, Hintz RL: A prospective, randomized study of testosterone treatment of constitutional delay of growth and development in male adolescents. Pediatr. 69: 681-7, 1982

323. Rosner W: Plasma steroid-binding proteins. Endocrinol.Metab.Clin.North Am. 20: 697-720, 1991

324. Rosner W: The functions of corticosteroid-binding globulin and sex hormone-binding globulin: recent advances. Endocr.Rev. 11: 80-91, 1990

325. Rubanyi GM, Freay AD, Kauser K: Vascular estrogen receptors and endothelium-derived nitric oxide production in the mouse aorta. J.Clin.Invest. 99: 2429-37, 1997

326. Rudman D, Drinka PJ, Wilson CR, et al: Relations of endogenous anabolic hormones and physical activity to bone mineral density and lean body mass in elderly men. Clin.Endocrinol. 40: 653-61, 1994

327. Salinas CA, Austin MA, Ostrander EO, Stanford JL. Polymorphisms in the androgen receptor and the prostate-specific antigen genes and prostate cancer risk. Prostate 65:58-65, 2005

328. Sandblom RE, Matsumoto AM, Schoene RB, et al: Obstructive sleep apnea syndrome induced by testosterone administration. N.Engl.J.Med. 308: 508-10, 1983

329. Santen RJ: Endocrine treatment of prostate cancer. J.Clin.Endocrinol.Metab. 75: 685-9, 1992

330. Santoro A, Viviani S, Zucali R: Comparative results and toxicity of MOPP vs ABVD combined with radiotherapy in PS IIB, III Hodgkin's disease. Proc.Am.Soc.Clin.Oncol. 2: 223, 1983

331. Sawaya ME, Shalita AR: Androgen receptor polymorphisms (CAG repeat lengths) in androgenetic alopecia, hirsutism, and acne. J Cutan Med Surg 3:9-15, 1998

332. Schaison G, Couzinet B: Percutaneous dihydrotestosterone treatment. In: Nieschlag E, Behre HJM: Testosterone - action, deficiency, substitution. Springer Verlag, Berlin, 423-36, 1998

333. Schänzer W: Metabolism of anabolic androgenic steroids. Clin.Chem. 42: 1001-20, 1996

334. Schiebel K, Weiss B, Wöhrle D, Rappold G: A human pseudoautosomal gene, ADP/ATP translocase, escapes X-inactivation whereas a homologue on Xq is subject to X-inactivation. Nature Genet. 3: 82-7, 1993

335. Schoch HK: The U.S. veterans administration cardiology drug-lipid study: an interim report. In: Holmes WL, Carlson LA, Paoletti R: Drugs Affecting Lipid Metabolism. Plenum Press, New York, 405-20, 1969

336. Schön M, Zaiac M, Schlag PM: Male breast cancer. Onkologie 18: 16-21, 1995

337. Schröder F: Does testosterone treatment increase the risk or induction of progression of occult cancer of the prostate? In: Bhasin S, Gabelnick HL, Spieler JM, Swerdloff RS, Wang C, Kelly C: Pharmacology, biology, and clinical applications of androgens. John Wiley, New York, 137-41, 1996

338. Schröder F: Impact of ethnic, nutritional, and environmental factors on prostate cancer. In: Bhasin S, Gabelnick HL, Spieler JM, Swerdloff RS, Wang C, Kelly C: Pharmacology, biology, and clinical applications of androgens. John Wiley, New York, 121-35, 1996

339. Schröder FH: Androgens and carcinoma of the prostate. In: Nieschlag E, Behre HJM: Testosterone. Springer Verlag, Berlin, 245-60, 1990

340. Schubert M, Bullmann C, Oettel M, Hübler D, Ernst M, Ezzidin S, Müller-Wieland D, Krone W, Jockenhövel F: Intramuscular application of testosterone undecanoate - a novel mode of androgen substitution with sustained duration. Exp.Clin.Endocrinol.Diabetes 107 (Suppl. 1): S62, 1999

341. Schubert M., Bullmann C., Minnemann T., Reiners C., Krone W., Jockenhövel F.: Osteoporosis in male hypogonadism: responses to androgen substitution differ among men with primary and secondary hypogonadism. Horm.Res. 60: 2 1-28, 2003

342. Schulte-Beerbühl M, Nieschlag E: Comparison of testosterone, dihydrotestosterone, luteinizing hormone, and follicle-stimulating hormone in serum after injection of testosterone enanthate or testosterone cypionate. Fertil.Steril. 33: 201-3, 1980

343. Schürmeyer T, Nieschlag E: Comparative pharmacokinetics of testosterone enanthate and testosterone cyclohexanecarboxylate as assessed by serum and salivary testosterone levels in normal men. Int.J.Androl. 7: 181-7, 1984

344. Schürmeyer T, Wickings EJ, Freischem CW, Nieschlag E: Saliva and serum testosterone following oral testosterone undecanoate administration in normal and hypogonadal men. Acta Endocrinol. 102: 456-62, 1983

345. Schweikert HU, Rulf W, Niederle N, Schäfer HE, Keck E, Krüskemper F: Testosterone metabolism in human bone. Acta Endocrinol. 95: 258-64, 1980

346. Seale TW, Flux M, Rennert OM: Reproductive defects in patients of both sexes with cystic fibrosis: a review. Ann.Clin.Lab.Sci. 15: 152-8, 1985

347. Seidell JC, Björntorp P, Sjöström L, Kvist H, Sannerstedt R: Visceral fat accumulation in men is positively associated with insulin, glucose, and C-peptide levels, but negatively with testosterone levels. Metabolism 39: 897-901, 1990

348. Seidman SN, Araujo AB, Roose SP, Devanand DP, Xie S, Cooper TB, McKinlay JB: Low testosterone levels in elderly men with dysthymic disorder. Am.J.Psych. 159: 456-9, 2002

349. Seminara SB, Hayes FJ, Crowley WF: Gonadotropin- releasing hormone deficiency in the human (idiopathic hypogonadotropic hypogonadism and Kallmann's syndrome): pathophysiological and genetic considerations. Endocr.Rev. 19: 521-39, 1998

350. Shabsigh R, Kaufman JM, Steidle C, Padma-Nathan H: Testosterone-replacement therapy with testosterone gel 1 % converts Sildenafil non-responders to responders in men with hypogonadism and erectile dysfunction who failed prior sildenafil therapy. Urol. 169 (Supple 247), 2003

351. Shabsigh R: Hypogonadism and erectile dysfunction: the role for testosterone therapy. Int.J.Impot.Res. Supple. 4: S9-13, 2003

352. Sheffer AL, Fearon DT, Austen KF: Clinical and biochemical effects of stanazolol therapy for hereditary angioedema. J.All.Clin.Immunol. 68: 181-7, 1981

353. Short KR, Nair KS: Mechanisms of sarcopenia of aging. J.Endocrinol.Invest. 22: 95-105, 1999

354. Signoretti S, Loda M: Estrogen receptor β in prostate cancer: Brake Pedal or Accelerator? Am.J.Pathol. 159: 13-6, 2001

355. Sih R, Morley JE, Kaiser FE, Perry HM, Patrick P, Ross C: Testosterone replacement in older hypogonadal men: a 12-month randomized controlled trial. J.Clin.Endocrinol.Metab. 82: 1661-7, 1997

356. Simon D, Preziosi P, Barrett-Connor E, Roger M, Saint-Paul M, Nahoul K, Papoz L. Interrelation between plasma testosterone and plasma insulin in healthy adult men: the Telecom Study. Diabetologia. 35:173-7, 1992

357. Simoni M, Gromoll J, Höppner W, Nieschlag E: Molecular pathophysiology of the pituitary gonadal axis. Adv.Exp.Med.Biol. 424: 89-97, 1997

358. Simoni M, Gromoll J, Nieschlag E: Molecular pathophysiology and clinical manifestations of gonadotropin receptor defects. Steroids 63: 288-93, 1998

359. Singh AB, Hsia S, Alaupovic P, Sinha-Hikim I, Woodhouse L, Buchanan TA, Shen R, Bross R, Berman N, Bhasin S. The effects of varying doses of T on insulin sensitivity, plasma lipids, apolipoproteins, and C-reactive protein in healthy young men. J Clin Endocrinol Metab. 87:136-43, 2002

360. Skakkebaek NE, Bancroft J, Davidson DW, Warner P: Androgen replacement with oral testosterone undecanoate in hypogonadal men: a double blind controlled study. Clin.Endocrinol. 14: 49-61, 1981

361. Slater S, Oliver RTD: Testosterone: its role in development of prostate cancer and potential risk from use as a hormone replacement therapy. Drugs and Aging 17: 431-9, 2000

362. Small M, McArdle BM, Lowe GDO, Forbes CD, Prentice CRM: The effect of intramuscular stanozolol on fibrinolysis and blood lipids. Thromb.Res. 28: 27-36, 1982

363. Smals AGH, Hermus ARM, Boers GHJ, Pieters GFFM, Benraad TJ, Kloppenborg PWC: Predictive value of luteinizing hormone releasing hormone (LHRH) bolus testing before and after 36-hour pulsatile LHRH administration in the differential diagnosis of constitutional delay of puberty and male hypogonadotropic hypogonadism. J.Clin.Endocrinol.Metab. 78: 602-8, 1994

364. Smith CG, Asch RH: Drug abuse and reproduction. Fertil.Steril. 48: 355-73, 1987

365. Smith EP, Boyd J, Frank GR, et al: Estrogen resistence caused by a mutation in the estrogen receptor gene in a man. N.Engl.J.Med. 331: 1056-61, 1994

366. Smith JC, Bennett S, Evans LM, Kynaston HG, Parmar M, Mason MD, Cockcroft JR, Scanlon MF, Davies JS. The effects of induced hypogonadism on arterial stiffness, body composition, and metabolic parameters in males with prostate cancer. J Clin Endocrinol Metab. 86:4261-7, 2001

367. Smith MR, Lee H, Nathan DM. Insulin sensitivity during combined androgen blockade for prostate cancer. J Clin Endocrinol Metab. 91:1305-8, 2006

368. Snyder PJ, Peachey H, Hannoush P, Berlin JA, Loh L, Holmes JH, Dlewati A, Staley J, Santanna J, Kapoor SC, Attie MF, Haddad JG, Strom BL: Effect of testosterone treatment on bone mineral density in men over 65 years of age. J.Clin.Endocrinol.Metab. 84: 1966-72, 1999

369. Snyder PJ, Peachey H, Hannoush P, Berlin JA, Loh L, Lenrow DA, Holmes JH, Dlewati A, Santanna J, Rosen CJ, Strom BL: Effect of testosterone treatment on body composition and muscle strength in men over 65 years of age. J.Clin.Endocrinol.Metab. 84: 2647-53, 1999

370. Soliman AT, Khadir MMA, Asfour M: Testosterone treatment in adolescent boys with constitutional delay of growth and development. Metabolism 44: 1013-5, 1995

371. Stanley HL, Schmitt BP, Poses RM, Deiss WP: Does hypogonadism contribute to the occurence of a minimal trauma hip fracture in elderly men. J.Am.Geriatr.Soc. 39: 766-71, 1991

372. Stellato RK, Feldman HA, Hamdy NAT, Horton ES, McKinlay JB: Testosterone, sex hormone-binding globulin and the development of type 2 diabetes in middle-aged men: prospective results from the Massachusetts male aging study. Diabetes Care 23: 490-4, 2000

373. Stuenkel CA, Dudley RE, Yen SSC: Sublingual administration of testosterone-hydroxylpropyl-beta-cyclodextrin inclusion complex simulates episodic androgen release in hypogonadal men. J.Clin.Endocrinol.Metab. 72: 1054-9, 1991

374. Sudhir K, Chou TM, Messina LM: Endothelial dysfunction in a man with disruptive mutation in oestrogen-receptor gene. Lancet 349: 1146-7, 1997

375. Sundaram K, Kumar N, Bardin CW: 7α-methyl-19-nortestosterone (MENT): an ideal androgen for replacement therapy. In: Bhasin S, Gabelnick HL, Spieler JM, Swerdloff RS, Wang C, Kelly C: Pharmacology, biology,

and clinical applications of androgens. John Wiley, New York, 493-97, 1996

376. Sundaram K, Kumar N, Bardin CW: 7α-methyl-19-nortestosterone: an ideal androgen for replacement therapy. Recent Prog.Horm.Res. 49: 373-7, 1994

377. Svartberg J, Midtby M, Bonaa KH, Sundsfjord J, Joakimsen RM, Jorde R. The associations of age, lifestyle factors and chronic disease with testosterone in men: the Tromso Study. Eur J Endocrinol. 149:145-52, 2003

378. Swerdloff RS, Heber D: Effects of aging on male reproductive function. In: Korenman SG: Endocrine aspects of aging. Elsevier Biomedical, New York, 119-35, 1982

379. Swerdloff RS, Jockenhövel F, Wang C: Endocrinology of the male. In: Kelley WN: Textbook of Internal Medicine. J.B. Lippincott Co., Philadelphia, 2013-21, 1992

380. Swerdloff RS, Wang C, Cunningham GR, Dobs AS, Iranmanesh A, Matsumoto AM, Snyder PJ, Weber T, Longstreth JA, Berman N: Long-term pharmacokinetics of transdermal testosterone gel in hypogonadal men. J.Clin.Endocrinol.Metab. 85: 4500-10, 2000

381. Takeda R, Ueda M: Pituitary-gonadal function in male patients with myotonic dystrophy - serum luteinizing hormone, follicle stimulating hormone and testosterone levels and histological damage of the testis. Acta Endocrinol. 84: 382-9, 1977

382. Tanaka S, Haji M, Nishi Y, Yanase T, Takayanagi R, Nawata H: Aromatase activity in human osteoblast-like osteosarcoma cells. Calcif.Tissue Int. 52: 107-9, 1993

383. Tayeb MT, Clark C, Murray GI, Sharp L, Haites NE, McLeod H: Length and somatic mosaicism of CAG and GGN repeats in the androgen receptor gene and the risk of prostate cancer in men with benign prostatic hyperplasia. Ann Saudi Med 24:21-6, 2004

384. Tenover JS: Androgen therapy in aging men. In: Bhasin S, Gabelnick HL, Spieler JM, Swerdloff RS, Wang C, Kelly C: Pharmacology, biology, and clinical applications of androgens. John Wiley, New York, 309-18, 1996

385. Tenover JS: Effects of testosterone supplementation in the aging male. J.Clin.Endocrinol.Metab. 75: 1092-8, 1992

386. The Endocine Society: 2nd annual consensus meeting on andropause. J.Clin.Endocrinol.Metab. http://www.endo-society.org/, 2001

387. Theilgaard A: Aggression and the XYY personality. Int.J.Law.Psychiatr. 6: 413-21, 1983

388. Thigpen AE, Davis DL, Milatovich A, et al: Molecular genetics of steroid 5α-reductase 2 deficiency. J.Clin.Invest. 90: 799-809, 1992

389. Thompson PD, Ahlberg AW, Moyna NM: Effect of intravenous testosterone on myocardial ischemia in men with coronary haert disease. Am.Heart J. 143: 249-55, 2002

390. Tibblin G, Adlerberth A, Lindstedt G, Bjorntorp P. The pituitary-gonadal axis and health in elderly men: a study of men born in 1913. Diabetes 45:1605-9, 1996

391. Tietz NW, Shuey DF, Wekstein DR: Laboratory values in fit aging individuals - sexagenarians through centenarians. Clin.Chem. 38: 1167-85, 1992

392. Tincello DG, Saunders PTK, Hodgins MB, et al: Correlation of clinical, endocrine and molecular abnormalities with in vivo responses to high dose testosterone in patients with partial androgen insensitivity syndrome. Clin.Endocrinol. 46: 497-506, 1998

393. Tricker R, Casaburi R, Storer TW, et al: The effects of supraphysiological doses of testosterone on angry behavior in healthy eugonadal men - a clinical research center study. J.Clin.Endocrinol.Metab. 81: 3754-8, 1996

394. Tsigos C, Latronico C, Chrousos GP: Luteinizing hormone resistance syndromes. Ann.N.Y.Acad.Sci. 816: 263-73, 1997

395. Tsitouras PD, Bulat T: The aging male reproductive system. Endocrinol.Metab.Clin.North Am. 24: 297-315, 1995

396. Tsitouras PD, Martin CE, Harman SM: Relationship of serum testosterone to sexual activity in healthy elderly men. J.Gerontol. 37: 288-93, 1982

397. Ulloa-Aguirre A, Blizzard RM, Garcia-Rubi E, et al: Testosterone and oxandrolone, a nonaromatizable androgen, specifically amplify the mass and rate of growth hormone (GH) secreted per burst without altering GH secretory burst duration or frequency or the GH half-life. J.Clin.Endocrinol.Metab. 71: 846-55, 1990

398. Umbreit K: Ist eine Substitution mit Estradiol auch bei Männern angezeigt? Gyne 14: 56-60, 1993

399. Urban RJ, Bodenburg YH, Gilkison C, et al: Testosterone administration to elderly men increases skeletal muscle strength and protein synthesis. Am.J.Physiol. 269: E820-6 1995

400. Valimaki VV, Piippo K, Valimaki S, Loyttyniemi E, Kontula K, Valimaki MJ. The relation of the XbaI and PvuII polymorphisms of the estrogen receptor gene and the CAG repeat polymorphism of the androgen receptor gene to peak bone mass and bone turnover rate among young healthy men. Osteoporos Int 16:1633-40, 2005

401. Van den Beld A, De Jong FH, Grobbee DE, Pols HAP, Lamberts SJW: Measures of bioavailable serum testosterone and estradiol and their relationships with muscle strength, bone density, and body composition in elderly men. J.Clin.Endocrinol.Metab. 85: 3276-82, 2000

402. van Kesteren P, Lips P, Gooren LJG, Asscheman H, Megens J: Long-term follow-up of bone mineral density and bone metabolism in transsexuals treated with cross-sex hormones. Clin.Endocrinol. 48: 347-54, 1998

403. van Kesteren PJM, Asscheman H, Megens JAJ, Gooren LJG: Mortality and morbidity in transsexual subjects treated with cross-sex hormones. Clin.Endocrinol. 47: 337-42, 1997

404. Veldhuis JA, Zwart AD, Mulligan T, Iranmanesh A: Muting of androgen negative feedback unveils impoverished gonadotropin-releasing hormone/luteinizing hormone secretory reactivity in healthy older men. J.Clin.Endocrinol.Metab. 86: 529-35, 2001

405. Veldhuis JA: Recent neuroendocrine facts of male reproductive aging. Exp.Gerontol. 36: 1281-308, 2000

406. Veldhuis JD, Urban RJ, Lizarralde G, Johnson ML, Iranmanesh A: Attenuation of luteinizing hormone secretory burst amplitude as a proximate basis for the hypoandrogenism of healthy aging in men. J.Clin.Endocrinol.Metab. 75: 707-13, 1992

407. Verhoeven G: Local control systems within the testis. Bailliere's Clin.Endocrinol.Metab. 6: 313-33, 1992

408. Vermeulen A, Kaufman JM, Goemaere S, van Pottelberg I: Estradiol in elderly men. Ageing Male 5: 98-102, 2002

409. Vermeulen A, Kaufman JM: Editorial: Role of the hypothalamo-pituitary function in the hypoandrogenism of healthy aging. J.Clin.Endocrinol.Metab. 75: 704-6, 1992

410. Vermeulen A, Rubens R, Verdonck L: Testosterone secretion and metabolism in male senescence. J.Clin.Endocr. 34: 730-5, 1972

411. Vermeulen A: Androgens and male senescence. In: Nieschlag E, Behre HJM: Testosterone. Springer Verlag, Berlin, 261-76, 1990

412. Vogel W, Klaiber EL, Broverman DM: A comparison of the antidepressant effect of a synthetic androgen (mesterolone) and amitryptilin in depressed men. J.Clin.Psychiatr. 46: 6-8, 1985

413. Vogt HJ, Borelli S, Heller WD: Fertilität bei Rauchern und Nichtrauchern. Z.Hautkr. 60: 1127-46, 1985

414. von Eckardstein S, Nieschlag E: Treatment of male hypogonadism with testosterone undecanoate injected at extended intervals of 12 weeks: a phase II study. J.Androl. 23: 419-25, 2002

415. Walsh S, Zmuda JM, Cauley JA, Shea PR, Metter EJ, Hurley BF, Ferrell RE, Roth SM.Androgen receptor CAG repeat polymorphism is associated with fat-free mass in men. J Appl Physiol. 98:132-7, 2005

416. Wang C, Alexander G, Berman N, et al: Testosterone replacement therapy improves mood in hypogonadal men - a clinical research center study. J.Clin.Endocrinol.Metab. 81: 3578-83, 1996

417. Wang C, Berman N, Longstreth JA, Chuapoco B, Hull L, Steiner B, Faulkner S, Dudley RE, Swerdloff RS: Pharmacokinetics of transdermal testosterone gel in hypogonadal men: application of gel at one site versus four sites: a general clinical research center study. J.Clin.Endocrinol.Metab. 85: 964-9, 2000

418. Wang C, Cunningham G, Dobs A, Iranmanesh A, Matsumoto AM, Snyder PJ, Weber T, Berman N, Hull L, Swerdloff RS. Long-term testosterone gel (AndroGel) treatment maintains beneficial effects on sexual function and mood, lean and fat mass, and bone mineral density in hypogonadal men. J Clin Endocrinol Metab. 89:2085-98, 2004

419. Wang C, Iranmanesh A, Berman N, et al: Comparative pharmacokinetics of three doses of percutaneous dihydrotestosterone gel in healthy elderly men - a clinical research center study. J.Clin.Endocrinol.Metab. 83: 2749-57, 1998

420. Wang C, Swerdloff RS, Iranmanesh A, Dobs AS, Snyder PJ, Cunningham GR, Matsumoto AM, Weber T, Berman N: Transdermal testosterone gel improves sexual function, mood, muscle strength, and body composition parameters in hypogonadal men. J.Clin.Endocrinol.Metab. 85: 2839-53, 2000

421. Wang C, Swerdloff RS, Iranmanesh A, Dobs AS, Snyder PJ, Cunningham GR, Matsumoto AM, Weber T, Berman N: Effects of transdermal testosterone gel on bone turnover markers and bone mineral density in hypogonadal men. Clin.Endocrinol. 54: 739-50, 2001

422. Wang C, Swerdloff RS: Evaluation of testicular function. Bailliere's Clin.Endocrinol.Metab. 6: 405-34, 1992

423. Wang YJ, Wu JC, Lee SD, Tsai YT, Lo KJ: Gonadal dysfunction and changes in sex hormones in postnecrotic cirrhotic men: a matched study with alcoholic cirrhotic men. Hepatogastroent. 38: 531-4, 1991

424. Webb CM, Adamson DL, Zeigler D, Collins P: Effect of acute testosterone administration on myocardial ischemia in men with coronary artery disease. Am.J.Cardiol. 83: 437-9, 1999

425. Webb CM, McNeill JG, Hayward CS, Zeigler D, Collins P: Effects of testosterone on coronary vasomotor regulation in men with coronary heart disease. Circulation 100: 1690-6, 1999

426. Weiss J, Axelrod L, Whitcomb RW, Harris PE, Crowley WF, Jameson JL: Hypogonadism caused by a single amino acid substitution in the β subunit of luteinizing hormone. N.Engl.J.Med. 326: 179-83, 1992

427. Wheeler MJ: The determination of bioavailable testosterone. Ann.Clin.Biochem. 32: 345-57, 1995

428. Wieacker PF, Knoke I, Jakubiczka S: Clinical and molecular aspects of androgen receptor defects. Exp.Clin.Endocrinol.Diabetes 106: 446-52, 1998

429. Wilson CA, di Clemente N, Ehrenfels C, et al: Müllerian Inhibiting Substance requires its N-terminal domain for maintenance of biological activity, a novel finding within the transforming growth factor-β superfamily. Mol.Endocrinol. 7: 247-57, 1993

430. World Health Organization Task Force on Methods For The Regulation of Male Fertility: Contraceptive efficacy of testosterone-induced azoospermia in normal men. Lancet 336: 955-9, 1990

431. World Health Organization Task Force on Methods For The Regulation of Male Fertility: Comparison of two

androgens plus depot-medroxyprogesterone acetate for suppression to azoospermia in Indonesian men. Fertil.Steril. 60: 1062-8, 1993

432. World Health Organization: Mesterolone and idiopathic male infertility: a double-blind study. Int.J.Androl. 12: 254-64, 1989

433. World Health Organization: WHO-Laborhandbuch zur Untersuchung des menschlichen Ejakulates und der Spermien-Zervikalschleim-Interaktion. Berlin, Heidelberg: Springer Verlag, 1993

434. Wright E, Fang J, Metter EJ, Partin AW, Landis P, Chan DW, Carter HB: Prostate specific antigen predicts the long-term risk of prostate enlargement: results from the Baltimore longitudinal study of aging. J.Urol. 167: 2484-7, 2002

435. Wu FCW: Testicular steroidogenesis and androgen use and abuse. Bailliere's Clin.Endocrinol.Metab. 6: 373-403, 1992

436. Yaffe K, Lui LY, Zmuda JM, Cauley J: Sex hormones and cognitive function in older men. J.Am.Geriatr.Soc. 50: 707-12, 2002

437. Yamamoto Y, Sofikitis N, Mio Y, Loutradis D, Kaponis A, Miyagawa I: Morphometric and cytogenetic characteristics of testicular germ cells and Sertoli cell secretory function in men with non-mosaic Klinefelter's syndrome. Hum.Reprod. 17: 886-96, 2002

438. Young NR, Baker HWG, Liu G, Seeman E: Body composition and muscle strength in healthy men recieving testosterone enanthate for contraception. J.Clin.Endocrinol.Metab. 77: 1028-32, 1993

439. Yusuf S, Hawken S, Ounpuu S, Dans T, Avezum A, Lanas F, McQueen M, Budaj A, Pais P, Varigos J, Lisheng L; INTERHEART Study Investigators. Effect of potentially modifiable risk factors associated with myocardial infarction in 52 countries (the INTERHEART study): case-control study. Lancet. 364:937-52, 2004

440. Zeegers MP, Kiemeney LA, Nieder AM, Ostrer H. How strong is the association between CAG and GGN repeat length polymorphisms in the androgen receptor gene and prostate cancer risk? Cancer Epidemiol Biomarkers Prev 13:1765-71, 2004

441. Zgliczynski S, Ossowski M, Slowinska-Srzdnicka J, Brzezinska A, Zgliczynski G, Soszynski P, Chotkowska E, Srzednicki M, Sadowski Z: Effect of testosterone replacement therapy on lipids and lipoproteins in hypogonadal elderly men. Atherosclerosis 121: 35-43, 1996

442. Zhang Y, Proenca R, Maffei M, Barone M, Leopold L, Friedman JM. Positional cloning of the mouse obese gene and its human homologue. Nature. 372:425-32, 1994

443. Zietz B, Schaffler A, Buttner R, Scholmerich J, Palitzsch KD. Elevated levels of leptin and insulin but not of TNF alpha are associated with hypertension in type 2 diabetic males. Exp Clin Endocrinol Diabetes 108:259-64, 2000

444. Zitzmann M, Brune M, Kornmann B, Gromoll J, Junker R, Nieschlag E. The CAG repeat polymorphism in the androgen receptor gene affects bone density and bone metabolism in healthy males. Clin Endocrinol 55:649-57, 2001

445. Zitzmann M, Brune M, Kornmann B, Gromoll J, von Eckardstein S, von Eckardstein A, Nieschlag E. The CAG repeat polymorphism in the AR gene affects high density lipoprotein cholesterol and arterial vasoreactivity. J Clin Endocrinol Metab 86:4867-73, 2001

446. Zitzmann M, Depenbusch M, Gromoll J, Nieschlag E. Prostate volume and growth in testosterone-substituted hypogonadal men are dependent on the CAG repeat polymorphism of the androgen receptor gene: a longitudinal pharmacogenetic study. J Clin Endocrinol Metab 88:2049-54, 2003

447. Zitzmann M, Gromoll J, von Eckardstein A, Nieschlag E. The CAG repeat polymorphism in the androgen receptor gene modulates body fat mass and serum concentrations of leptin and insulin in men. Diabetologia 46:31-9, 2003

448. Zitzmann M, Nieschlag E. The CAG repeat polymorphism within the androgen receptor gene and maleness. Int J Androl 26:76-83, 2003

Index

Index